Edward Atkinson

The Science of Nutrition

Edward Atkinson

The Science of Nutrition

ISBN/EAN: 9783742817808

Manufactured in Europe, USA, Canada, Australia, Japa

Cover: Foto ©Klaus-Uwe Gerhardt /pixelio.de

Manufactured and distributed by brebook publishing software (www.brebook.com)

Edward Atkinson

The Science of Nutrition

THE SCIENCE OF NUTRITION.

TREATISE UPON THE SCIENCE OF NUTRITION,
BY EDWARD ATKINSON, LL. D., PH. D.

THE ALADDIN OVEN,
INVENTED BY EDWARD ATKINSON.
WHAT IT IS. WHAT IT DOES. HOW IT DOES IT.

DIETARIES CAREFULLY COMPUTED
UNDER THE DIRECTION OF MRS. ELLEN H. RICHARDS.

TESTS OF THE SLOW METHODS OF COOKING IN THE ALADDIN OVEN,
BY MRS. MARY H. ABEL AND MISS MARIA DANIELL,
WITH INSTRUCTIONS AND RECIPES.

NUTRITIVE VALUES OF FOOD MATERIALS,
COLLATED FROM THE WRITINGS OF PROF. W. O. ATWATER.

APPENDIX:
LETTERS AND REPORTS.

SPRINGFIELD, MASS., U. S. A.
CLARK W. BRYAN & COMPANY, PRINTERS.
1892.

COPYRIGHT, 1892, BY EDWARD ATKINSON.

INTRODUCTION.

It had been my intention to adopt a more modest name for this compendium of my experiments, and I am warranted only by the approval and advice of my scientific friends and co-laborers in a field that has yet been only partially explored in adopting the form of my title. Without their aid and counsel I should have been able to give only a few crude hints for others to work out.

Much delay has occurred in the publication of this book. It has grown upon my hands, and in its final shape it is not in as good a form as I could wish; but time is wanting for further correction. As the line of investigation has opened, it has become apparent that we were at the very beginning of what will be necessary in order to establish a true Science of Nutrition; yet more before the facts that have been developed by science may be converted into a simple art that any one may practice. Suffice it that by the aid of my friends I may have been able to put, in this unsatisfactory form, a considerable part of what may yet be called the alphabet of the Science of Nutrition. I venture to hope that others more competent may take up this work and perfect my crude inventions.

In one of my earlier treatises upon the "Art of Cooking" I ventured to bring out the importance of this subject in the following terms:

[POPULAR SCIENCE MONTHLY, NOVEMBER, 1889.]

I will challenge attention and discussion by first submitting some very positive and dogmatic statements, subsequently sustaining them by such proofs as I have to offer:

1. Special apparatus for broiling and frying has been adequately and suitably developed for the use of those who can afford these somewhat wasteful methods of preparing food, yet excellent when skillfully practiced.

2. The ordinary methods of frying are utterly bad and wasteful.

3. Bread may be baked suitably in a brick oven and also economically when the work is done upon a large scale.

4. It is very difficult to bake bread in a suitable way in the common iron stove or range; for this, among other reasons, most of the bread consumed in this country is very bad, although we have the greatest abundance of the best material.

5. Meats may be well roasted, in a costly manner, before an open fire.

6. Aside from the exceptional apparatus or methods named, substantially all the modern cooking stoves and ranges are wasteful and more or less unsuitable for use. All the ordinary methods of *quick* baking, roasting and boiling are bad; and, finally, almost the whole of the coal or oil used in cooking is wasted.

7. The smell of cooking in the ordinary way gives evidence of waste of flavor as well as a waste of nutritious properties; in most cases the unpleasant smell also gives evidence that the food is being converted into an unwholesome condition, conducive to indigestion and dyspepsia.

8. Nine-tenths of the time devoted to watching the process of cooking is wasted; the heat and discomfort of the room in which the cooking is done are evidence of worse than waste.

9. The warming of the room or house with the apparatus used for cooking is inconsistent with the best method of cooking and might be compassed at much less cost if the process of cooking were separated from the process of warming the room or dwelling.

10. No fuel which cannot be wholly consumed is really fit to use in the process of cooking.

The true science of cooking consists in the regulated and controlled application of heat, by which flavors are developed and the work of converting raw and indigestible material into nutritious food is accomplished.

For this purpose the quantity of fuel required is almost absurdly small compared to the quantity commonly used.

In establishing the standard of heat for different processes or different kinds of food I have found it rather difficult to measure the temperature of the oven in its everyday work, and I have also broken a good many high-grade thermometers in experimenting. I have one very useful cooking thermometer, which is made by Joseph Davis & Co., Fitzroy Works, London, S. E., but I can find nothing corresponding to it in this country.

A simple method in experimenting is to place a small vessel containing lard oil on the same shelf with the food which is in process. Take the heat, on the completion of the work, by immersing the bulb of the thermometer in the hot fat.

I have also procured a large number of alloys or metals possessing different melting points, of which I place small pieces alongside or inside the cooking vessels in small saucers, and in this way I can make a close approximation to the maximum heat. Once fairly established, the matter becomes simple by gauging the size of the burner to the size of the oven.

I was led into my recent experiments in the art of cooking by dealing with the statistics of the cost of food to the multitude. It comes to one-half the cost of living or more, to at least nine-tenths of the people of this country and of every other country. The less the income the greater the proportion spent for food.

My immediate incentive in trying to solve the practical problem was my happening to be present, on a cold day, when a number of workmen in a mill which they were building were about to take their dinner. When they opened their dinner pails a mess of cold victuals was disclosed, which seemed to me must require the digestive power of an ostrich to dispose of. This led me to take up again a line of experiments upon which I had begun many years before, but had dropped. I had tried to make a cooking-pail which could be carried in the hand, and in which the workman could cook his dinner without any attention being required while he was at work. I have now accomplished this, but the way has been long, devious, wasteful and costly and the device is not yet in a commercial form.

It may be observed that the progress of almost all invention is from the simple, rude and laborious ways of the untaught,—through very complex and costly devices, ending at last in the simplest methods of the most effective and least arduous kinds. So it has been in the making of cooking utensils and in the art of cooking. So it has been with my own efforts upon these lines.

The final conclusions in which I have summed up the results of my experience may be laid down in the following very simple propositions, which any one can apply who knows how or who is willing to be taught. It is oftener the will than the capacity that is lacking.

1st. The simple rule for making a cooking apparatus is as follows:

Take a paper box, then take some heat from a lamp, put it into the box and keep it there. When you have enough heat in the box, having first put the food which is to be cooked into suitable pots or pans, put them into the box with the heat; keep them there until the heat and the food are combined; keep up the supply of heat from the lamp. (Aladdin Oven Patent 524,949.)

2nd. The simple rule for learning how to prepare and how to cook food, is as follows:

Take one part of gumption and one part of food; combine them together in a dish in which the food may be cooked; put that dish, as directed by rule No. 1, into the box where the heat is; keep it in there long enough to cook the food.

There is no mystery and no difficulty about either of these matters. Any boy or girl can be taught in one lesson how to master certain simple principles so as to become a good plain cook, after sufficient practice—*provided* each already possesses the one part of gumption which must be combined with the food. All the cookery books and all the recipe books that I have ever read wholly omit this necessary ingredient. Those who do not yet know how to apply these two rules, may read this pamphlet; some may perhaps profit by it.

I am told that it may be injudicious to claim so much, or to make so light of the difficulties in training a cook. That may be true; but, what can one do? I must either state the facts as they are, or suppress them. I have tested these two rules in practice, and I have not failed. The boys that I selected to experiment upon possessed the element of gumption, and they worked very well on the first trial.

When correcting the proofs of this text, the rules which I have laid down for making an oven and educating a cook suddenly took on an appearance of grotesque absurdity. Then I bethought me that any reader might say to himself, "What does this man mean? Does he himself more than half believe what he says?" And to that question my answer might be, "*I* doubt if *he* does."

When we recall the fact that men of the highest scientific attainments have tried to solve this problem at various times, and have not found out the way, it really does seem rather absurd that my two rules should suffice. For instance, we may recall the work of the French inventor Papin, who, more than two hundred years ago, while engaged in the apparently more important work of laying the foundation for the modern steam-engine and while improving the air-pump, yet found time to invent the Papin soup-digester. That utensil remains in use under the inventor's name, down to the present day, a most durable monument. It is a most excellent utensil, but it is adapted only to the single purpose of utilizing soup stock, leaving nothing but the bones. It does not suffice for common use with us, because a great many of our working people declare that they do not want "bone soup," and they stigmatize simmered food of the best kind under the name of "pig wash."

Passing down over a century, the next name which at once comes up is that of Benjamin Thompson—Count Rumford. He devoted a lifetime to the study of heat, and he believed that by the invention of the Rumford oven he had solved the very problem on which we are still engaged. Yet the Rumford oven has gone almost wholly out of use, and Count Rumford did *not* accomplish his purpose, because he had not a cheap fuel which could be regulated and completely consumed in the process of heating the oven. Mineral oil had not been discovered and even coal was not much in household use. In other words, the true science of cooking waited for the invention of methods of making gas, for the discovery of kerosene oil and for the conversion of wood into wood pulp or "indurated fibre."

Again, consider the constant effort of inventors in the improvement of the iron cooking range and the iron cooking stove. Witness their mis-directed efforts to ventilate ovens and draw off the vapor generated in the oven from the food; which is precisely what ought not to be done when the heat is under proper regulation. Witness their efforts to enable every one to cook food quickly, which is absolutely the wrong way in nine times out of ten. Witness all the other devices which have been applied to the perfection of the iron stove; yet the common iron stove remains an infernal machine, almost unfit to be used, even by the intelligent, in any true method of cooking, and incapable of being applied to any really scientific process of cooking.*

When we recall all these instances and efforts, it really does seem absurd that any one should claim to have found out a way by which a boy or girl, in whose training a little gumption has been developed and who never cooked anything before, can be taught in one hour the simple principles, which after sufficient practice will enable them to take some heat from the top of a lamp, to put it into a box and then to subject all kinds of meat, game, poultry, fish, fruit, grain and vegetables to this heat in such a way, that the most skillful "Chef" or "Cordon Blue" will fail by comparison in developing the finest natural flavor of each distinct variety of food. It may seem absurd; nevertheless, it seems to be a fact.

I can only explain this case by taking over to myself a story which is told about the invention of what is called the California pump; one of the simplest devices for the rough work of draining. It consists of a slanting trough in which revolves a leather belt, across which belt strips of wood are nailed. When the man who first adapted this pump to draining a ditch showed it to one of his neighbors, never thinking that there was anything very wonderful about it, his neighbor's comment was this: "Why, Jim, that is so simple that nobody but a fool would ever have thought of it." That is about the aspect in which the Aladdin Oven and the Workman's Pail are presented to my *alter ego*. When my double takes himself a little way off, so as to be able to look back on what the other fellow is trying to do; or in order to comprehend what his practical representative is occupied about in the material work of mundane existence; he sometimes puts the question to me,

---
[From the Nation, No. 1361.]

* "If we consider that wonderful work of human hands, the kitchen range, under the management of the regular cook, who knows how to put on all the draught at once and keep it on, what a devourer of fuel it is! We need a cup of tea or a chop in summer, and a fire is kindled that would generate steam enough to drive an ocean racer a mile upon her course, the kitchen is turned into a Tophet, the miserable servants swelter in the apartments which their own stupidity and that of mankind have rendered uninhabitable, and their employers are rendered uncomfortable above. The extravagance of the Chinese, who, as related by Charles Lamb, at first thought it necessary to burn down a house whenever they wanted to roast a pig, is nothing to ours. In place of all these trials, heating gas is now supplied to those who have the wit to use it and the enterprise to supply themselves with gas ranges, which can be lighted and extinguished in a moment."

or I put it to him, or we put it to both of us: "Do you really suppose that you have accomplished the work which Count Rumford undertook, and did not bring into practical form?" It remains for others and not for me or my double to answer this question.

It has taken me a long time to put myself in a position to have the question asked, but it has lately been answered by a cloud of witnesses.

In dealing with the preparation of food I have been led to give some consideration to the possible increase in the variety and in the quantity of our food supply. I will treat only one or two points in this connection—it opens a broad field.

To the uninstructed mind it hardly seems as if our scientists had given sufficient attention to the nutrition either of the animal or the plant in the utilization of the leguminous plants classed as pulse, *i.e.*, pease and beans in great variety. In a recent investigation (in 1888) of the wheat supply of the world, I was much struck with the statement made by Dr. George Watts, Reporter on Economic Products to the Government of India, that the limit of exports of wheat from India had probably been reached, because no more land within easy reach of railways or waterways could be spared from the cultivation of rice and *pulse*. Rice alone, consisting too much of starch, would not give vigor; the pulse furnishes the necessary nitrogen. The farmers of the Southern states, who cannot produce ripened corn for milling in competition with the West, can yet produce corn stalks for the silo in excessive abundance. In their cottonseed meal they have a fat-producing food also full of phosphates, and in the cow-pea vine they have a plant of almost unequalled power in renovating the soil when plowed under; also of the utmost value when mixed with corn stalks in the silo, making a complete food, even for working cattle and mules.

I also infer from this report of Dr. Watts that we yet know but little in this country about many varieties of rice and of pulse. Prof. Church states that "rice is more largely grown and consumed as food than any other cereal. Alone, however, it is not a perfect food, being deficient in albuminoids and in mineral matters." In speaking, however, of the food of the Japanese, he remarks that "both swamp rice and *mountain rice*, when grown in Japan, contain a higher percentage of albuminoids than is usual with this grain." What is mountain rice?

In respect to one variety of pulse, the Soy or Soja bean, Prof. Church remarks: "The Soy bean is entitled to the highest place, even amongst the pulses, as a food capable of supplementing the deficiencies of rice or other starchy grains. Very few vegetable products are so rich as this bean, at once in albuminoids and in fat and oil, the former constituent averaging 35 per cent. and the latter 18 or 19."

At the time of the Cotton Exposition, in Atlanta, 1881, I imported from China two or three bushels of Soy beans, which were distributed from there. I have lately seen several references to their cultivation in Southern agricultural journals. The sugar planters of Louisiana have only within very recent years discovered the fertilizing properties of cottonseed meal; the Chinese have used bean meal as a fertilizer for centuries.

Dr. Watts remarks in the report to which I have referred: "There are rices that can be cultivated *on comparatively dry soils,* rices that occupy the ground more than half the year, *rices that grow in cold, temperate climates,* and rices that can thrive only in tropical swamps."

Speaking of pulse, he says: "The pulses alone occupy more ground than either wheat or rice." He also gives a list of the principal varieties, as follows: "Gram or Chicken pea, Dal or Thur, Mung, Urad, Moth, Shim, Poput, Kulthi." "The Dal is a large woody shrub sown around the margins of the fields, or in rows through wheat and cotton. It does not, therefore, interfere with wheat cultivation. All the other pulses or leguminous crops are procumbent herbs."

If another great crop of cotton should so reduce the price as to force a variation in Southern farming with a little more rapidity, it might prove to be a blessing in disguise. Would it not be judicious to procure the seed of mountain rice from Japan, of upland rice from India, and might it not be judicious to add to the Soja of China the Dal, Poput and Gram from the northwestern or temperate sections of India?

The last time I was at Columbia, S. C., I was told that the Pea Ridge farmers, were thriving, who had learned to nourish the poor soil of the ridges, from which the original fertility of the land had been washed down into the bottom lands, by turning under the Cow-pea vines, while the planters on the bottoms, whose life in old times used to be so easy, were no longer prospering. Would it not be judicious for Governor Tillman to send to India for some "Dal," and would it not be prudent for the Southern Farmers' Alliance to let our silver go to India in exchange for some "Poput" and "Gram," instead of keeping it here to debase our credit?

Prof. Atwater tells us that nitrogen is the most important and most costly element in the nutrition of man, as it also is in the nutrition of the soil.

Here is a great class of nitrogenous pulses waiting to be imported. Perhaps they are no better than our native cow-pea vines, but who knows? What do we know about English horse-beans or Indian pulse? Why should Englishmen enjoy "*Stachys Tubifera*" from Japan while we have it not? A new tuber might give as pleasant a sensation as a new lily.

The science of nutrition leads one very far afield, and unless I stop here on this subject the introduction may become longer than the main treatise.

CONCLUSION.

I have not hesitated to take the somewhat unusual methods which I have adopted in preparing and publishing this pamphlet, in order to assure the wide circulation of the theories, the facts and the figures which are printed herein. A true economy in the use of food and the attention which is demanded by the art of nutrition have failed until very lately to receive the attention that is due to their importance.

Unless one is prepared to spend a small fortune in the ordinary modes of advertisement, or unless one invents some new method of securing public attention by exciting curiosity, as I have attempted to do in this pamphlet, the spread of information on any given subject is about as slow and difficult as it is to accomplish the object aimed at in the processes about which the information is sought to be given. I have known most useful and valuable inventions to be delayed for many years for want of the means, the capacity or push necessary to force them into public notice; some have been delayed so long that the inventors have died poor; one known to me even committed suicide in despair. I doubt not many valuable devices have been wholly lost for the time being, both to the inventor and to the public, on this account.

How to avoid the ways of the vendors of quack medicines, and at the same time to secure public attention, is a difficult problem; yet more difficult when the subject of notice is one of which the vendor desires to promote the sale for other purposes than personal profit, or to which he cannot give the time or attention which would be necessary in order to make a sufficient profit corresponding to his effort. It is not my purpose to go into an extensive manufacture of ovens on my own account, unless I am obliged to do it lest the possible benefit of my work should be lost. If my ovens are justified in use, they may soon be taken up by those who are more conversant with the business than I am; then my own profit or royalty on the patent, if any, may be devoted to the further development of the science of nutrition.

EDWARD ATKINSON

PART I.

TREATISE UPON

THE SCIENCE OF NUTRITION.

SUBSTANCE OF A LECTURE DELIVERED IN THE LAW LECTURE ROOM OF COLUMBIA COLLEGE, NEW YORK, AT THE INSTANCE OF PROF. THOMAS EGLESTON—SUBSEQUENTLY AMENDED AND COMPLETED.

Ladies and Gentlemen:
Before presenting my subject, permit me to say that what I have written is not intended for publication as a conclusive statement. It is wholly tentative and is incomplete.

If this treatise should be printed in its present form I must append to it the *caveat* to which I was accustomed when I earned my living as an accountant: "E. & O. E."—"Errors and Omissions Excepted."

You may be learning the alphabet while I have got as far as words of three letters in the study of the subject which is to be dealt with to-day.

What I shall exhibit to you in the product of these ovens and cooking pails will speak for itself, and may be fully reported according to the facts.

I understand these meetings to have been arranged by Prof. Egleston in order that I might bring before you some evidences of the progress that I have made in applying scientific methods to practice in the most necessary of all the arts, this art of nutrition. Yet I am not a scientist, therefore all my own work is empirical, tentative and crude. It merely opens the way into a field where true science needs to be applied, but which is as yet almost unoccupied.

Since I have ventured almost alone to work in this field upon an entirely new method, you will pardon me if I am obliged to speak in the first person singular.

At first I found it very difficult to overcome the sense of its being quite unfit for me to speak to women about the art of cooking, or to physicians upon the science of nutrition. I have, however, put aside that feeling since I have discovered what my mission in the world is. I suppose every one has a mission, whether conscious or unconscious of it. My own mission appears to be to overcome the *inertia* of woman; a very hard piece of work.

Five or six years since the names of *proteids, albuminoids* and *carbo-hydrates* conveyed no meaning to my mind except as having something to do with raising fat cattle and feeding pigs.

I have, however, a propensity for statistics, although no one values figures less or distrusts them more than I do, unless I can wrest from them their true meaning. In some of my social studies the figures disclosed to me the rather appalling fact, namely,—that the price of life to about nine-tenths of the people of this land of abundance comes to one-half or more of their incomes; the supply of food being the source of material life. In other words, half the cost of living or more is spent for food material by the vast majority of the people of this and every other land. When I fully comprehended what these figures meant, my interest in the victualling department became aroused.

At about the time, some five years since, when these figures had interested me, after a long rest from any practical work in the food question, I happened to be passing over a new factory then in process of construction, which was to be insured in the Factory Mutual Insurance Company, of which I am president. It was the noon hour and the workmen were opening their dinner pails. When I saw what was in them my interest in the victualling department took on a practical direction. I made up my mind that I would invent a cooking pail in which a workman might carry his food to the place of his work, cook it after he got there, and have a hot and nutritious dinner at 12 o'clock. Incomplete and somewhat clumsy as these pails which are before you may be, I am very far on toward the accomplishment of my purpose. All inventions help each other. I shall soon be in possession of a new burner, which is already invented for a kerosene lamp, by means of which such perfect combustion of kerosene oil may be assured as perhaps to require no glass chimney. It will need only a perforated plate for a wind guard. I shall then be able to attach a platform to the bottom of my pail on which I may place a small lamp which will do the cooking in the two cooking boxes, which can be put inside the pail, each containing one pound of food; one day a meat stew in one of the boxes, oatmeal in the other; another day, a fish chowder in one, and an apple dumpling in the other.

In the use of these pails the question is no longer with me, "What can I do?" but "What can I not do?" The work is almost incredible that may be done in this simple device. When the lamp and pail are all in one piece, that can be carried in one hand, less cumbrous than even this one, my mission to the workman will have been accomplished. He may find a suitable place, either in doors or out of doors, to hang up the pail, where it will be out of the way of sharp winds; light the lamp, and at noon have his hot, nutritious and perfectly digestible meal ready for him to partake of.

In my progress toward this objective point I have, according to the common custom, worked by devious ways. I have wasted a good deal of money and a good deal of time, mainly because I am not a master of any science and because my practice is wholly empirical. In my profession I am, however, obliged to take note of the progress of almost all the sciences, and the rule is about the same; through devious and complex methods to ultimate simplicity. Yet more: I am very often called upon to take note of the many things which science has not yet accomplished, or has only partly accomplished. In the little lamp in which the complete combustion of a small quantity of liquid fuel may be accomplished without a chimney, I may find evidence that I have not been far wrong in a remark that I have often made about the combustion of fuel under boilers for the purpose of making steam, to wit: that even when we have applied all the knowledge that science has yet imparted to us, our ignorance of the true method of converting coal into power may be measured by the height of our chimneys and by the strength of the draught or blast that we must apply in order to keep up the combustion of the coal. I think gaseous fuel must soon come into almost universal use, for smelting, generating steam, and for cooking also.

If that remark has any approach to being true in respect to the combustion of fuel under the modern boiler, how much greater our ignorance must be in the application of coal to the processes of cooking. If a small part of the paper upon which this treatise is written were compressed and made use of in the way that coal is now consumed in the modern marine boiler to drive the engine of the steamship, this paper would serve to generate force enough to drive much more than a ton of food and its proportion of a freight steamer two miles upon the sea, and yet in the best marine boilers and engines I believe less than twelve per cent. of the potential of the fuel consumed is converted into work. Yet such is our ignorance of the right methods of applying heat to the cooking of food in our common iron stoves or ranges, that we burn more pounds of coal than the number of pounds of food to which the heat is to be applied. If the common range or stove now in position in your dwellings were used only for heating the kitchen, boiling water, boiling potatoes and heating water for circulation through the house, the saving of the mere *excess* of coal which is now burned in order to force the oven to a heat suitable for quick cooking would weigh more than the food to which that heat is applied. I think, but I am not sure on this point, that one-half or more of the coal which is now used to do the cooking, as well as to warm the kitchen and heat water, may be saved by the adoption of my apparatus for the cooking, while only depending on the ordinary range or stove for the rest of the service. That is the testimony of my cook in my own winter house, where my kitchen is very large and is on the north side, on the top of a hill in the country. At my summer place by the sea-side I have not had a pound of coal on the premises for three years. We burn a little wood in the cooking stove for special purposes, but the greater part of our work is done in my ovens.

Yet this saving of fuel is the matter of the very least relative importance, if I may safely accept the conclusions which others have reached on the testimony of now nearly two hundred witnesses. It almost required this great mass of testimony to convince myself of the facts in this case, and to give me the necessary confidence to appear before you and others.

It is the waste of food and the conversion of good food into bad feeding that is the motive of my work and of this address.

I may perhaps venture to name this treatise a "Sermon upon the Potato Gospel," borrowing that phrase from Carlyle. If it is a sermon of serious purport, as I think it may be entitled to be called, I may divide it into the customary three parts of a discourse :

    1st.   The Selection of Food.
    2nd.   The Preparation of Food.
    3rd.   The Application of Heat to its Chemical Conversion.

As to the last title, one may perhaps ask, Why use three long words and five short ones? Why not say "cooking" at once, and have done with it? I will presently give you good reasons for my choice of this phrase.

In regard to the selection of food. If it were a question of feeding horses, cows or pigs, all the necessary information could be found in almost innumerable popular treatises, magazines and agricultural papers. Any one can learn in a day how much and what to give to a trotting horse ; how much and what to feed to a working horse ; how much and what to put before a milch cow or a pair of oxen ; but if one of you had asked, only a little while since, how to select the right ingredients and proportions of food for the nutrition of men and women, you could only be referred to some abstruse and scientific treatises. Even yet the more popular treatises of Prof. Atwater, of Prof. Church, of Sir Henry Thompson, of Mrs. Ellen H. Richards and others, are apparently but little known ; while Mrs. Abel's " Lomb prize essay" upon cooking is the only American cookery book known to me in which any exact instructions are given for working with a measured heat. If there are any cookery books except French, known to any one among my hearers, in which the heat that is to be applied is defined in anything but general terms,—such as a slow oven or a quick oven,—I should be glad to have the title of the work.

Again, every one who has anything to do with horses or cows knows the difference between oats and corn ; between horse hay and rowen. He may even know that the *proteids* or nitrogenous elements of the oats are required to maintain the muscular energy of the horse. He may know that the fat in Indian corn, cracked, can be assimilated by the working horse, generating greater but slower force. He may know that the full proportion of starch (carbo-hydrates) must go either with the nitrogen or fat in due proportion. To how many has it ever occurred that there should be as definite instructions for feeding men and women as there now are for feeding horses and cows? The food of men must be suitable to the kind of work and to the intensity of the work—so that it may be assimilated. Each human being must have the true proportion of *proteid*, of *fat* and of *starch*, with lesser proportions of the mineral salts; else, even with an abundance, he may not be well nourished. True, that through a process of natural selection, and by way of experience without scientific knowledge, each race and each nation has found out the kind of food or the combinations that will give the right proportions of nutrients at the least cost. But as a rule the food of men and women is served without the slightest attention to proportions or to waste, following only, under the pressure of necessity, a sort of blind instinct. I do not propose to bring each man or each woman to a measured quantity every day. That would be very foolish to undertake; but may there not be certain broad and general rules, which, when once laid down, may serve to give direction to the purchase of food material, thereby assuring full nutrition with the saving of that vast waste which is almost a disgrace to this nation?

The chemical standards of nutrition which have been established by Professor Voit and others in Germany, by Sir Lyon Playfair in England, by Dr. Pavey, and other competent authorities, vary in some measure from the American standard. What I name the American standard is that which has been elaborated mainly through the investigations of Professor William O. Atwater, Mrs. Ellen H. Richards, and others. It contains a somewhat larger proportion of fat than the European dietaries; perhaps thereby becoming more suitable to the colder and more changeable conditions of the climate of the northern section of the United States, and to the more energetic life of our people.

Prof. Atwater has converted the chemical units of nutrition in all these various dietaries into units of heat or *Calories*. At this standard they are all substantially alike even though they vary in some measure in their chemical elements. It is possible that the professor has builded better than he knew.

We may now adopt the Calory as the unit of nutrition. It may then become a very simple matter to prepare rules and tables that shall be a true guide to intelligent persons in the purchase and in the consumption of food; not day by day, but by the adoption of standards corresponding in every way to the chemical elements and to the units of heat, say for thirty days. I think we may even put the whole art of nutrition by and by, into the common school arithmetics, in the form of examples of addition, multiplication and the like; it seems to me that they would be very much better lessons for children than many of the logical puzzles in figures which I have found in the school arithmetics that are wholly unfit to be there; perplexing children instead of teaching them.

I have made some progress in this matter, and I will hereafter submit some of the results in the form of tables. They have not yet been revised by competent authorities to the full extent, but are in a broad and general way consistent with the true standards of nutrition.

The Calory or mechanical unit of heat may well be adopted as the unit of capacity for work either mental or manual—either with hand or brain. In its application to food material it works its conversion into nutritious food, and in the further conversion of the food by assimilation to the sustenance of the body it is the synonym for healthy existence, strength and activity. To the direction of this force, and to the end that it may make and not mar wholesome living, attention is now being given.

The exact standard of nutrition for a man at active but not excessive work is 700 grams of actual nutritive and digestible material free of water; 450 of carbohydrates or starch; 150 of fats; 150 of protein; with such mineral ingredients as will be found in any miscellaneous dietary in sufficient measure. These elements will yield 3,520 Calories or mechanical units of heat, the Calory being the amount of heat necessary to raise one kilo or 1,000 grams of water one degree centigrade. Of course persons vary in the quantity of food which can be assimilated, according to many varying conditions of life.

In order to make allowance for unavoidable waste we may safely adopt 4,000 Calories as the average *units of nutrition* for a man at active but not excessive work for one day.

On this unit we may make variations by percentage in ratio to the kind of work done, the sex and the weight of the consumer.

In about the ordinary proportions of grain, meat and vegetables in which food is purchased, I find that one pound containing the proportions of starch, fat and nitrogen required by the American standard yields about 1,200 Calories.

A day's ration of 4,000 Calories therefore calls for three and one-third to three and a half pounds of food material of the ordinary kinds. Tea, coffee and the sugar and cream used therein, water or other liquids consumed as beverages, not being included in this computation. Beverages, except cocoa, possess but little food value.

Now, while it might be unreasonable to expect an exact measure or unit of nutrition to be adopted and put in practice day by day, it becomes a very simple

THE SCIENCE OF NUTRITION.   17

matter to establish a rule for the purchase of 100 pounds of food per month, or thirty days' rations at three and one-third to three and a half pounds each.

As a tentative measure yet to be more accurately computed, I have made the following table consistently with those general rules.

It is not a fancy table. In order to get a true basis for the retail prices of the cheaper kinds of food in Boston, some of which, notably the prices of potatoes and of course hominy or samp, are very high, I employed a lady who is much interested in this matter to get prices in Boston at the South End, the superintendent of a branch of the New England Kitchen at the North End, the colored cook who is employed in my office kitchen at the West End, and my office boy, a bright lad, to get prices in the neighborhood of our largest market—Quincy Market. I averaged these four returns in making this table, and I have printed them all in one of my circulars.

My first table is made on the basis of the cheaper kinds of meat and fish and the best kinds of flour, grain and vegetables, all bought at retail. There is a considerable margin for reduction if these articles were bought in large quantities. If bread is bought rather than baked at home on my methods, the price of bread taken at two and a half cents a pound must be doubled in Boston. In New York you can buy better bread at three cents a pound cash at Mr. Samuel Howe's National Bakeries than we can get in Boston at six cents.

In reducing the pounds of food to nutrients and Calories I have assumed that the meat and vegetables will be purchased in variety and I have therefore taken the average of each class of foods in my computation.

Thirty days' rations, yielding substantially 1,200 Calories per pound, in the proportion of three elements of starch to one of protein and one of fat.

Milk may be substituted for some of the meat fat or pork, but of course in much larger measure by weight.

See subsequent tables given in Part II, which have been carefully corrected since this lecture was given.

Table No. 1 is now given as an example. This is given in place of a less accurate dietary which was given in the original lecture.

18   THE SCIENCE OF NUTRITION

DIETARY NO. 1, FOR 30 DAYS.

Suitable in nutrients (protein, starch and fats) and in Calories or mechanical equivalents of heat, to the full nutrition of an adult, occupied in work which gives moderate exercise. Price of flour by the sack or barrel—all other prices for small quantities at retail.

CONSTANTS SUITABLE TO DIETARIES 1 TO 12, INCLUSIVE.

22 lbs.	Flour at 2½ cts.		.55
3 "	Oatmeal, at 4 cts.		.12
3 "	Cornmeal, at 3 cts.		.09
6 "	Hominy, at 4½ cts.		.27
2 "	Butter, at 28 cts.		.56
2 "	Suet, at 6 cts.		.12
10 "	Potatoes, at 2½ cts.		.25
2 "	Sugar, at 5 cts.		.10
3 "	Cabbage, at 3 cts		.09
2 "	Carrots, at 2½ cts.		.05
2 "	Onions, at 5½ cts.		.11
57		2.31	Calories, 79,770

CONSTANTS IN QUANTITY, VARIABLE IN PRICE ACCORDING TO KIND AND QUALITY.

Low priced.

12 lbs.	Beef, neck or shin, at 6 cts.	.72	
5 "	Mutton, neck, at 6 cts.	.30	
4 "	Bacon, at 12 cts.	.48	
2 "	Beef liver, at 6 cts.	.12	
1 "	Veal, at 8 cts.	.08	
1 "	Pork, at 8 cts.	.08	
25		1.78	Calories, 24,256
82 lbs. total, 30 days,		$4.09	104,026
2 73-100 lbs. for one day,		.136	3,467
Cost per week,		.952	

With less meat and fat and more vegetables and milk the weight will be greater.

The exact standards of American nutrition are much higher than those of Europe, as they may well be in order that the much higher rates of wages which our people earn may be fully justified and sustained by the greater amount of potential energy which our abundant product of food enables us to supply at low cost. They are as follows:

THE SCIENCE OF NUTRITION.

		Exact.	With Waste Added.
No. 1.	Man at hard work,	4,060	Calories, 4,600
" 2.	Man at moderate work,	3,520	" 4,000
" 3.	Man at light exercise or woman at moderate work,	2,815	" 3,200
" 4.	Woman at light exercise,	2,300	" 2,600

On this basis, on the cost of the foregoing standard, Class 1 would require daily:

3 7-8, say 4 lbs. food, about 1,200 Calories per lb., at 5 cents per lb., 20 cts.
Class 2.
3 3-8, say 3 1 2 lbs. food, about 1,200 Calories per lb., at 5 cents per lb., 17 1-2 cts.
Class 3.
2 5-8, say 2 3-4 lbs. food, about 1,200 Calories per lb., at 5 cents per lb., 13 3-4 cts.
Class 4.
2 1 8, say 2 1 4 lbs. food, about 1,200 Calories per lb., at 5 cents per lb., 12 1-2 cts.

On a minimum basis, therefore, yet one which may be readily adopted, the nutritive material which is necessary for a man at hard work in Boston can be purchased in small quantities at retail prices at 20 cents a day or $1.40 per week.

For a man at moderate work, at 17¼ cents a day or $1.22½ per week.

For a man at light exercise or a woman at moderate work, at 13¾ cents a day or .96¼ per week.

For a woman at light exercise, at 12½ cents a day or .87½ per week.

As these purchases would be made in pounds, and as in every element a suitable addition has been made for what may be called reasonable waste, the common measure would be substantially as follows:

PER DAY.

Meats,	about	10	oz.
Suet or Fat,	"	1	"
Salt Pork,	"	1	"
Butter,	"	1 1-2	"
Fish,	"	3	"
Bread,	"	14	"
Hominy, Oat and Corn Meal,	"	6	"
Beans or Pease,	"	2	"
Sugar,	"	3	"
Roots and other Vegetables,	"	12 1-2	" 54 ounces.

That does not sound like a very meagre diet.

If eggs are added at four per week at 3 cents each (city prices for fresh eggs), .12 cts.
With fruit at per week, .09 "
Tea and Coffee, .07 "

Four cents per day, .28 cts.

the unit of nutrition at the low priced standard would be as follows:

PER DAY.

Class 1, Required Food, 20   Add Eggs, Fruit and Beverage at 4 cts., .24
" 2,   "   " 17 1-2   "   "   "   "   "   at 4 " .21 1-2
" 3,   "   " 13 3-4   "   "   "   "   "   at 4 " .17 3-4
" 4,   "   " 12 1-2   "   "   "   "   "   at 4 " .16 1-2

PER WEEK.

Class 1, $1.88   including Eggs, Fruit and Beverages.
" 2,   1.50 1-2   "   "   "   "   "
" 3,   1.24 1-2   "   "   "   "   "
" 4,   1.15 1-2   "   "   "   "   "

I am not yet prepared to say how much this ration No. 1 would weigh with the water added in the processes of cooking.

The bread is computed with the water in it which the flour takes up at 40 per cent. on the weight of flour.

Meats and fish may be combined with water in different portions. As the coarse or tough parts are better in soups, stews and hashes, the water added would be in large proportion.

Hominy and meal take up several times their weight in water, while vegetables shrink both in the preparation and in cooking. I should think this unit of nutrition at three and one-third pounds would weigh about five pounds after cooking. I am too much of a sedentary man, and I find that my average ration of cooked food with water taken up in the process, aside from beverages, is about three and a half pounds, and on the basis of the dietary submitted I could live well at $1 per week.

In anticipation of this meeting I devoted Fast Day, April 2d, to some experiments. I caused one of my ovens to be substantially filled with food, in eight combinations. This experiment is repeated in one of these ovens to-night, which we will presently open.

THE SCIENCE OF NUTRITION.

The food purchased was as follows, at Brookline prices:

MEAT AND FISH.

2 lbs.	Shin Beef, without bone, 7 cents,	. . .	.14	
2 "	Veal, " " 5 cents,	. . .	.10	
2 "	Beef Liver,	. . .	.16	
1-2 "	Bacon,	. . .	.07	
2 "	Mutton flank,	. . .	.10	
1 "	Suet,	. . .	.05	
2 "	Halibut Nape,	. . .	.10	
1-2 "	Salt Pork,	. . .	.06	
12 lbs.	Meat Fat and Fish,	. . .		.78

GRAIN.

1 lb.	Hominy,	. . .	.04
3-4 "	Oatmeal,	. . .	.03
1 "	Corn Meal,	. . .	.03
1-4 "	Rice,	. . .	.02
3 lbs.		.	.12

VEGETABLES.

4 lbs. Potatoes, . . .	.	.10
1 " Turnips, onions, etc.,	. .	.05
5 lbs.		.15
4 lbs., 2 quarts, Skimmed milk, .	. .	.16
Salt, Spice, etc., say . .	. .	.04
20 lbs. solid,	. . . .	$1.25
4 " liquid,		
12 " water,		
36 lbs.		

Put into the oven at 9.30 a. m.
Removed from oven at 2 p. m.
Weight of cooked food, 32 lbs.
Loss by evaporation of water, 4 lbs.

I have not computed the Calories or nutrients in this quantity. It is intended to show what can be done in one oven at one time. It took somewhat less than one hour to prepare the eight combinations, namely one dish each: Beef and hominy hash; veal, hominy and tomato; mutton stew; liver and bacon with corn meal; halibut nape and potato, baked; oatmeal, plain; poor man's pudding; baked Indian pudding.

You may presently test the quality of these combinations. We have put in forty pounds this time and may take out thirty-five to thirty-six.

With such a lot of food, strong in fat and nitrogen, there should be at least twelve pounds of bread and twelve pounds of vegetables, say twenty-four pounds at two and a half cents=.60.

24 lbs. added, ...	.60
36 lbs. cooked as recited, . . .	1.25
Add for dried apple sauce or some cheap kind of fruit,	.15
Total, 60 lbs., at a cost of	$2.00

You don't half believe it. Nor did I until I had proved it more than once. Presently you shall have an object lesson when this oven is opened.

Now, if my methods of combining nutrition and arithmetic by the standard of Calories and nutrients has no other result, it may lead to as much attention being given to the nutrition of the human beings in city institutions as is given to feeding the horses in the city stables. From some communications that have been made to me I am inclined to think that right at this point is a place for scientific charity.

I gave a seven-course dinner party at my house a few days since to my whist club and friends, including oranges and coffee, which cost thirteen cents each for food material. Each cigar consumed after dinner cost more than the dinner.

I lately gave a dinner of four courses, soup, fish, meat and vegetables, and mush with molasses for dessert, to nine of the poorer students at Harvard who want to economize, and there were three others. Each had a pound and a half of strong food. The cost for the twelve was sixty-one cents.

The quantity of oil of 150 degree flash test that will be consumed on this forty pounds of food will be one and a half pints at one and a half cents per pint, two and a quarter cents.

There may be some variations to be made on the American standard.

In the judgment of Sir Henry Thompson, with whom I consulted personally, we concentrate our food a great deal too much in this country; we fine down our flour and lose in nutrition as well as in bulk; we also eat too much meat and fat. Moreover, the ration of meat and fat which may serve us well in winter may serve us ill in summer.

In order to meet these variations we may change the thirty day dietary. When we substitute more grain and vegetable food, we reduce the Calories per pound, and we therefore require more pounds a day.

We may also change the nutrients from 3 starch to 1 protein and 1 fat to four starch, 1.1 protein and 1 fat.

In Part II tables one to twelve, inclusive, will serve as a true guide in this matter, but the proportions must be subject to variation by personal experience. for all. (See revisions in subsequent tables.) There is no hard and fast rule.

It will be observed that in these low-priced dietaries we have not included eggs. In our factory boarding-houses in Massachusetts, the consumption of eggs per adult is one every other day. One egg every other day, at 16¼ cents a dozen, comes to $3 a year per adult. We may compute our present population as being equal to 50,000,000 adults; 50,000,000 adults, at $3 each, would spend $150,000,000 a year for eggs.

The factory boarding-house standard cannot be declared a very extravagant one, as "the Mealers," who come from a distance to their work, are supplied and served with twenty-one meals per week, at $1.60 to $1.75 for women; $2.00 to $2.25 for men.

I know of no other place where such fairly well-cooked meals are furnished at so small a charge. Of course there is no rent charged to this account.

This is a sketch of the elements of the science or art of nutrition which may perhaps be perfected and may possibly be taught in the common school arithmetics.

I have thus presented a theory of nutrition at a minimum cost. But it could only be attained in practice under our existing conditions by people of more than common intelligence. It would be almost impossible to attain with the use of the cooking apparatus now in common use; yet at this standard, witness to what incomprehensible figures we are led when dealing with the whole mass of our population.

Whoever is right as to the enumeration of the census year, we now number about 65,000,000 in 1891. If we make the utmost reduction for children of ten or under, keeping in consideration the larger need of growing children from ten to seventeen; also bearing in mind the great proportion of all who get their living by "the sweat of their brow" and who must have ample nutrition, we cannot estimate the consumption of food at less than what would be required by 50,000,000 adult men at active or moderate work or women at active work, corresponding in their requirements to a standard of nutrition somewhat under Class 2 and little over Class 3, or 14¼ cents a day, which comes to one dollar a week for food; to which we may add a quarter of a dollar a week for eggs or milk, tea, coffee and fruit.

At $52 per year the food necessary to sustain 50,000,000 adults comes to	$2,600,000,000
At $13 per year, the eggs or milk, tea, coffee and fruit come to . .	650,000,000
Food and wholesome beverages, ...	$3,250,000,000
The most conservative estimate of the cost of beer, wine and spirits to the consumers is	750,000,000
Total, ...	$4,000,000,000

The problem with which we are dealing to-night in the mere alphabet, we are at the mere beginning of an attempt to apply the same science to the nutrition of man, that we have so long been attempting to apply to the nutrition of the soil, the plant and the beast of the field; our problem is how to cook not less than

THREE BILLION DOLLARS' WORTH OF MEAT, FISH, GRAIN AND VEGETABLES A YEAR.

I think you will admit, ladies and gentlemen, that it was quite time to apply science in the kitchen and to develop a true and simple art of cooking.

At what price will you measure the waste of labor, the waste of fuel, the waste of heat, the waste of comfort, the waste of temper, the waste of health, the waste of morality and sobriety due to the waste upon whisky which is again due to lack of well-cooked food? Is this waste a billion dollars' worth of potential energy a year? Is it not a great deal more?

Perhaps you will come to the conclusion that the potato gospel will bear a great deal of preaching.

Ought there not to be a cooking laboratory attached to every agricultural experiment station? Would one be out of place even in Columbia College?

But, ladies and gentlemen, there is another aspect of this case. True economy does not consist in living on shin of beef or halibut nape. Very few people can afford to waste the time that would be required to live at a cost of a dollar a week, or at fourteen cents a day, unless obliged to do so. There is a certain horse-sense in the reply of the workman to some of these suggestions when he says, "We don't want your bone-soup" and "We won't have your pig-wash. Give us something better than what we are used to, or better cooked than we now get it."

Suppose we double the price, put up the prices of the animal food to rates which would be paid for good solid meat free of bone and of good quality; add a pint of milk, an excess of butter and one egg every day. I think this ration corresponds closely to the average consumption and expenditure in the families of well-to-do people; if we add to this food ten cents a day for tea, coffee, condiments and fruit, we make the total equal to fifty cents a day for each inmate of the household.

It is probable that much more than half a pound of clear meat free of bone is brought into the kitchen of well-to-do people who are not extravagant, for each adult every day. I have made a few inquiries at some of the very best hotels—and am satisfied that there are many in which the price paid for food per day is one dollar per inmate or two dollars per guest, counting one person in the service of the house to each guest.

THE SCIENCE OF NUTRITION.

TABLE 2, COMPUTED AT FIFTY CENTS PER HEAD INCLUDING EXTRAS.

Four pounds food at about 1,200 Calories per pound = 4,800 Calories, one pound milk added, 310 Calories, making five pounds = 5,110 Calories.

1-2 lb.	Clear Meat, at an average of 20 cts. per lb.,					.10
1-4 "	Fish,	"	"	16	" "	.04
1-4 "	Butter,	"	"	36	" "	.09
	1 Egg,	"	"	36	" per doz.,	.03
1	" 1 Pint Milk,					.04
1	" Bread,					.03
1-4 "	Oatmeal, rice, hominy or corn meal,					.01
1 1-4 "	Vegetables,					.03
1-2 "	Sugar,					.03
5 lbs.						.40 cts.
	Tea, Coffee, Fruit and Spices extra.					.10 "

This would be an excessive allowance for a man at very hard work, if it were all cooked in such a way that it could be assimilated; with water added it would weigh as much as seven pounds.

You will find that fifty cents' worth a day per inmate comes into your houses, even if you think you live rather simply. It cannot be consumed. How much is wasted?

I myself eat about three and a half pounds a day, including soup, but not including fruit or beverages.

At my standard about three-fifths of that ration, at forty cents, or twenty-four cents' worth, suffices.

If proper methods of cooking were adopted and right methods of utilizing what is now wasted in the ordinary method of buying and preparing this food material, all the potential energy in this dietary could be enjoyed by the average adult, at twenty-eight cents a day or in round figures at two dollars per week.

At two dollars a week the food bill of 50,000,000 adults comes to $5,200,000,000 a year.

By statistical analysis of our crops and food products, projected from the gross valuations and wholesale data to their points of ultimate consumption, I am fully satisfied that the food bill of this country is in fact as much as five billions of dollars and is probably more.

I am also satisfied that one-fifth part of this huge volume of good material is converted *into bad feeding.* In other words, for lack of science in the economy of the kitchen $1,000,000,000 worth a year of potential energy is wasted. How shall we save it?

THE SCIENCE OF NUTRITION.

Life is but a conversion of force which takes the form of clothing, food and shelter. That is all that any of us get out of the material products by which we are sustained, sheltered and clothed.

If this misdirection of force of one billion dollars' worth were converted by the saving of that part of the potential energy in the food consumed which is now wasted—or if this force were given a new direction into the work that would be required to provide house-room—nearly twice the shelter now enjoyed would ensue. The worst problem with which we are called to deal would then be solved—the housing of the masses.

The family now in one room could have two; the destruction of the poor is their poverty, and the worst waste of food is among them. The family now in three rooms, if once taught the art of nutrition, might have five, and those who now have as many rooms as are required could direct the force now wasted to greater comfort and to the higher plane which only in the long run makes life worth living.

Have my figures led you away into visionary conceptions, such as any one may find in the dry columns of statistical science if only he has the eye to see what is written between the lines or is inscribed behind the columns?

Let us then get back at once to ovens and to the practical problem of food and feeding. I have expended as much money as I can afford in my own empirical but somewhat practical methods of developing this oven. My good friends, Mr. Andrew Carnegie, his partner Mr. Phipps, and several of my Boston friends have furnished a part of the means by which Mrs. Richards has been enabled to work out the scientific application of right methods of applying heat to some of the processes of cooking. A very benevolent lady, whose life is full of good works, has furnished the capital for establishing the New England Kitchen, through which the products of my process as well as of other processes of plain, wholesome cooking are being distributed on a commercial basis, in the shape of wholesome, nutritious and well-cooked food at moderate prices which yield a reasonable profit.

I have now engaged a competent teacher of cooking, who, under the general direction of Mrs. Richards, will work out the beginning of the problem of a cooking laboratory. As I have stated, it is in this matter that I may move faster the better I am sustained. I shall devote my own profits in the oven business to the oven development as far as they may go. I do not desire any assistance from any one in that part of my undertaking, but the development of a cooking laboratory opens a much wider field and it is beyond me.*

---

* Subsequently to this lecture Mr. Theodore A. Havemeyer placed $1,000 at my disposal for further research and $6,000 at the disposal of Professor Egleston for the establishment in New York of a kitchen corresponding to the New England Kitchen of Boston.

We will now come to the second part of my discourse—*the preparation of food.* After having filled a shelf in my library full of cookery books, to which I have given careful attention, adding treatises on nutrition and physiological essays on alimentation, I have come to a very simple conclusion covering the whole ground. The two simple rules for qualifying any one after a little practice to become a good plain cook and first-rate bread maker may be repeated.

Take a little heat off the top of a small lamp; put it into a wooden box with a little water in order to raise your bread; take some heat off the top of a large lamp, put it into a wood-pulp or paper box, and keep it there to bake your bread or to cook your food. That is the first rule.

In the preparation of food the single rule is this: Take one part of gumption and one part of food; mix them together in a dish or pan, then put the compound into the oven and keep it there long enough for the heat to do the work. Practice will soon tell each one how long.

It is not a very difficult matter, for instance, to make bread. I thought it had been until I tried. Then I found that bread-making in the common way, kneading by hand, took a good deal of muscle and not much mind; but I do not fancy paws and perspiration in my bread; the idea is unpleasant even to speak of. After my first lesson I therefore taught my cook how to mix the dough with a stout wooden spoon. There is nothing new in the method. I now find that a few people had learned this secret before, but not many. The work can be done just as well and more quickly. My objection to the ordinary method was well expressed in a story lately told me by an old gentleman who took his little grandson to the bakery. The next day the boy wanted to go again, and gave this reason: "I want to see that fat man who was washing his hands in the brown bread. He gave me a cooky." We now use the Stanyan bread-kneader, which is a sort of mechanical spoon and knife combined; it is much easier to work than the wooden spoon.

I then found that there was apt to be a good deal of trouble, delay and uncertainty in raising the dough. I therefore bought a Case Bread-Raiser, which assures the exact measure of fermentation in three and a half to four hours. With these three bits of apparatus, the kneader, the raiser and the oven, any one who can read can learn to make good bread in half an hour so as never to miss it. Baking takes two hours in the Aladdin Oven already heated, or a longer period if you like a thick, tender crust and tawny color, with nutty flavor all through the loaf, due to the partial change of the starch into dextrine or grape sugar. Here are examples of bread which no human hand has touched even from the time the wheat was planted until it was taken from the pan in which the loaf was baked.

But suppose we lay out a dinner. (See subsequent instructions.)

28   THE SCIENCE OF NUTRITION.

I told my office boy one day to prepare just a dinner of five courses for ten people. He had never cooked anything before. He went and did it. Every one said it was a good dinner, and I thought so myself. That kind of a dinner is in one of these ovens.

If the true object in cooking is to develop and retain all the fine natural flavor of meat, fish, vegetable, fruit and grain; if the right method is one that will prevent the more volatile portions of the fats and juices being carried off in bad smells; if you desire that the food after it is cooked shall be in a perfectly nutritious and digestible condition, the oven indicates the way to attain all these objects. There is no constant watching, no stirring the materials over a hot stove to prevent scorching or burning. If you desire sauces or gravy they can be prepared over another lamp or in a chafing dish. There are no heavy hods of coal to be lifted. There is no dust. There are no ashes to be removed. One quart of oil burning eight hours in one of these lamps gives off heat enough from the top of the chimney to cook fifty or sixty pounds of bread, meat and vegetables, in three separate charges. In one of these ovens now before you there are forty pounds of eight different kinds of food, which, when served, will have had expended upon them one pint and a half of oil valued at about two cents.

I have observed that in nearly all the cookery books or recipe books the chief part of the rules laid down are for making combinations of food material; rules for the proportions of each; the quantity of meat and other material; the pounds of flour; the dozens of eggs; the ounces of sugar and spice and the way to mix them.

Of course, any person of common intelligence can follow these printed rules for such combinations.

I next observe that almost all the other instructions in the books are mainly directed to overcoming the defects in the common stoves and ranges. You are told how to stir the compounds in order to prevent scorching or burning; how the pan must first be placed at a given point and just brought to a simmer without being allowed to boil, and then moved to another part of the stove where the material will only simmer. You are told that for some things you must have a hot oven or a quick oven; for others a slow oven; but no accurate definition is given as to what "quick" or "slow" may happen to be.

All these variations in the directions and in the processes may be attributed to the irregularity and to the constant high heat which is due to the continuous blast and to the method of combustion that is necessary to keep up any roasting or baking heat in the oven of the iron range or stove. You are told that you must not accumulate vapors in the oven, because they may be noxious. This is all nonsense. If the oven is subjected to the right method of heat the vapor of the food itself is the best surrounding that you can give it. But when you crack or dissociate fats, then it is necessary to remove the vapor in order to avoid the unpleasant flavor which burnt fat may give to what is being cooked. I can place and have often placed meat, fish, cauliflower, onions, tomatoes and custard pudding in the same oven at the same time. Neither imparts any flavor to the other, because I only evaporate a little water from either kind, without distilling or dissociating the juices or the fats. How many times I have heard ladies exclaim, "How I wish I could teach my cook how to simmer anything. Why will they always keep such a roaring fire?"

My first invention was what might have been called a hot-water oven; an apparatus in which one could only simmer and stew. I tried to introduce it first among those who need instruction most, but I have come to the conclusion that the only way to help the ignorant poor is first to ameliorate the condition of the rich.

To speak seriously, the true method of developing this subject is to deal neither with rich nor poor, but with those who possess intelligence and who desire to avoid waste. I have lent out twelve or more cooking-pails; some of them to ladies who are teachers, who have been in the habit of getting their own breakfast and dining at restaurants; some of them to students in Harvard University who are under the necessity of working through at the least cost; some of them to ladies of my acquaintance who are eager to try experiments for the benefit of others, and so on.

The next question which I propose to solve in my present series of experiments and in my cooking laboratory, under the direction of an experienced teacher of cooking, will be the making of sauces and flavors that may correspond to or bring out yet more fully the natural flavors of the solid food with which they are served. I think that the greater part of the complexity of the customary rules and methods for working in this department may be done away with. I think that where one has a sure and certain control over the source of heat, the making of fine sauces, which is one of the secrets of the professed cooks, may become as simple as the cooking of the food itself. This is not fancy work. It is a true part of the science of nutrition. In spite of the gross waste of the multitude there is reason to believe that many persons, especially among the more cultivated classes of the community, are not sufficiently nourished because they try to live too simply, or do not choose the right kinds of food. We may also find women, especially those who serve in shops, who live on very poor food and drink any quantity of tea because they do not know how to prepare good food or how to tempt the palate by making it appetizing. I think it is an important element in the science of nutrition to overcome these depraved tastes by tempting the palate with appetizing sauces that can induce these classes of people to eat more nutritious food. These are merely impressions which I have gained from my empirical studies. Probably many of you among my hearers are more competent to pass upon this point than I am myself.

If, then, the absolute control of the source and the measure of heat is so important, are we not led directly to the question, "In what does good cooking consist?" And that question brings me to the third part of my discourse. It consists in the application of heat to the chemical conversion of food material into nutritious and appetizing food. It is a somewhat difficult scientific problem to bring into the form of rules, but I am convinced that when these rules are established the art may be made very simple, and may be readily taught to any one of ordinary comprehension and aptitude.

What is the function of heat in its application to food? My scientific friends must give the chemical statement of the work. In its practical result is it not to convert the hard grain into nutritious bread?—the raw, sapid, tough and tasteless fibers of meat into tender and appetizing dishes?—to render it possible to get the infusion of coffee from the roasted berry?

Now, when you bear in mind the messes that are so often put before you—the bad bread, the hard, indigestible and offensive products of the American frying-pan and all the other abominations that come out of the kitchen—you may begin to appreciate the importance of holding complete control over the source of the heat so that it can be regulated.

What intelligent control can any one exert over the heat derived from anthracite coal in an iron stove or range? What does the average cook know about what is going on in the oven?

Is not by far the greater part of the heat derived from the fuel which is either forced by the strong draft up the chimney or else radiated into the room? I wonder what part of the potential of the coal is converted into actual work in our common processes of cooking?

When we put instruments which are scarcely fit to be used by the most skillful cooks into the hands of untrained and uninstructed persons and demand of them a fine process of chemical conversion, what right have we to expect any better results than what we get?

It seems to me most strange that it should have remained for myself to devise an oven made of non-metallic and non-heat conducting material in which heat may be encased at any established degree. Why have not ovens always been made in this way?

I have kept heat for a long time at over 400° F., in a small oven of which the paper or wood-pulp wall was only five-eighths of an inch thick, yet I could bear my hand anywhere upon the outer case.

May I not then claim the following results in my processes of conversion?

1st. Digestibility. The long application of a very moderate degree of heat makes all tough meats tender.

2nd. Although I add nothing to the nutrients which are in the food, I avoid much evaporation of the contained water. Is not the food more digestible as well as more appetizing in consequence of this retention of all its natural juices and nearly all the water that is in it when in the uncooked condition?

3rd. By keeping the heat below the dissociating or "cracking" point of the animal fats, do I not keep the fat in a condition to be assimilated in the process of digestion? Whereas, is it not true that a high heat which drives off the volatile portion of the fat leaves the remainder in an indigestible condition?

4th. That my processes do most fully preserve and develop the fine, natural flavors of grain, fruit, meat, fish and vegetables is now so fully proved by the testimony of very many witnesses that this point may no longer be put in the form of a question.

For a long time I suspected my own imagination of misleading me: but one lady has written me that last summer she *tasted* for the first time many vegetables which she had been eating all her life.

Neither is it any longer necessary for me to put questions about the relief from care, attention and discomfort, or from dust and ashes; here is the testimony, not yet of a thousand, but of more than a hundred witnesses on these points.

There is one aspect of this question which gives me great satisfaction.

It opens the way to the re-establishment of the unit of the family as the unit of society and of civilization.

Our success in adopting the collective method in our great factories and workshops, in which after all only about ten per cent. of our working population are occupied: our success in solving the problem of moving a year's supply of meat and bread a thousand miles at a cost to the common laborer of a day's wages: our success in diminishing the margin of profit on the railway service to so low a point that if the New York Central could get the present value of the empty barrel at the end of the haul of a barrel of flour a thousand miles it would pay ten per cent. in place of five, has put a glamour over a great many otherwise sensible people.

Even some who claim to be economists want to adopt the collective method or to extend it over all the conditions of society. Hence the rush to the legislature and to Congress to remedy all the ills of life by legislation. One man would abolish poverty by piling all the taxes on land; another by nationalizing industry, whatever that may mean; another by stopping importations and making more work with less product; another by destroying all the tenement houses, as if no house at all were better than some kind of a house when it is overcrowded and dirty, and so on throughout the gamut.

It seems to me that we shall never elevate humanity until we teach men and women how to live. We shall never make homes happy except we work on the unit of the family.

If I can show how a family of five can be nourished as well and as cheaply as a phalanstery of five hundred then I may claim that in developing the simple art of nutrition and in doing away with what one of my correspondents calls "the slavery of stove" I may deserve a pleasant remembrance in the homes of the future.

I am now prepared to present my object lessons and to answer any questions if they are not too scientific.

(Subsequently to the delivery of the lecture, the contents of the five ovens and four pails were served to about two hundred people.)

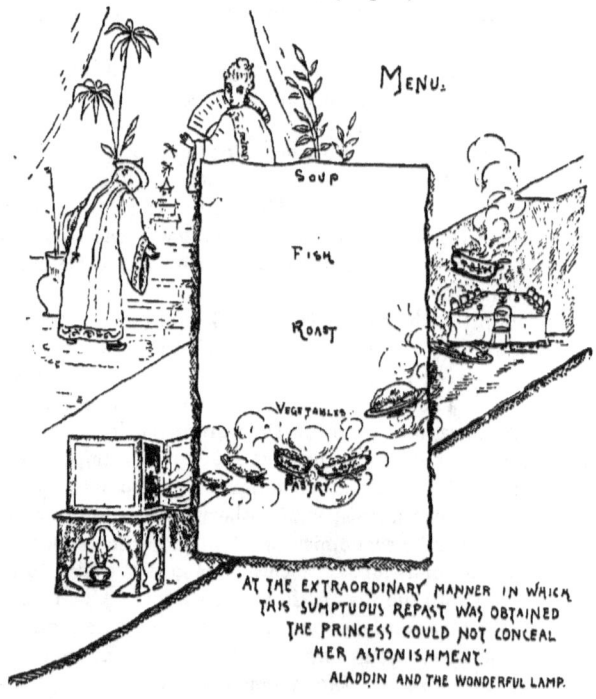

"AT THE EXTRAORDINARY MANNER IN WHICH THIS SUMPTUOUS REPAST WAS OBTAINED THE PRINCESS COULD NOT CONCEAL HER ASTONISHMENT."
ALADDIN AND THE WONDERFUL LAMP.

FOOD POWER.

We may regard food as the element of life upon which the power of man to sustain himself wholly depends. Anything else can be spared, food cannot be. Again, if the measure or quantity of food is not sufficient and is not rightly adjusted to the conditions of complete nutrition, both the manual and mental efficiency of man will be impaired. If the force which is generated by the assimilation of food is not adequate to the complete support of the workman, he will become incapable of making the product out of which his wages are paid in sufficient measure to re-supply himself for the subsequent efforts necessary for the maintenance of his productive power. Life is a conversion of force, and if the force or food power supplied to the man is inadequate or incomplete, the right conversion of force cannot go on.

It will be found that the rates of wages in different countries vary as the supply of food is abundant or scanty. As a matter of fact we find that the rate of wages is highest, or in other words the effective earning or productive power of man is greatest in the United States, where the supply of food is most abundant. Australia and New Zealand may also be named as places where a small population is able to produce for itself the most abundant supply of food and where the wages or earnings are very high.

Next in the abundance of the food supply comes Great Britain, and in Great Britain again we find the highest rate of wages outside of the United States and the British Colonies already named.

Next in order follow Holland, France, Germany, Belgium, Italy, Austria, and Russia in Europe. In each of these countries, in the order named, the supply of food lessens as compared to the other; either because the land is poor or is cultivated without the best mechanism, or because a very large share of the product is diverted for the support of great armies so that the remainder becomes insufficient.

Finally, when we pass into Asia we find the supply of food only equal to a bare subsistence and correspondingly very low rates of wages prevail. How far one factor may be taken as an antecedent or consequent of the other will not be discussed in this treatise.

This reference to the very far-reaching nature of the food question with which we are dealing, is made only as an introduction to the following tables and statements which have been prepared and selected by Mrs. Ellen H. Richards in order to show: First, why food is required; Second, how much is needed; Third, of what it should consist; Fourth, at what cost it can be procured; and Fifth, how to tell when enough food of the right varieties has been consumed.

Mrs. Richards deals with the subject in the following terms, incorporating tables from different sources:

The animal body is a living machine, capable of doing work,—raising weights, pulling loads, and the like. The work of this kind which it does can be measured by the same standard as the work of any machine, i. e. by the mechanical unit of energy—the foot-ton. The power to do mechanical work comes from the consumption of fuel, the burning of wood, coal or gas; and this potential energy of fuel is often expressed in units of heat or *Calories*,* a Calorie being nearly the amount of heat required to raise two quarts of water *one* degree Fahrenheit. The animal body also requires its fuel, namely food, in order to do its work or its thinking or even its worrying.

But the animal body is more than a machine, it requires not only fuel to enable it to work, but to *live* even without working. About one-third of the food eaten goes to maintain its life, for while the inanimate machine is sent periodically to the repair-shop, the living machine must do its own repairing day by day and minute by minute. Hence it is that the estimations of the fuel and repair-material needed to keep the living animal body in good working and thinking condition are, in the present state of our knowledge somewhat empirical, but it is believed that within certain wide limits useful calculations can be made by any one willing to give a little time and thought to the subject. Our knowledge may be rapidly increased if such study is made in many localities and under many varying circumstances.

To afford a stimulus to such study the following tables are given as illustrations only of what may be found in the various works on the chemistry of food and nutrition. The student is especially referred to the articles by Professor W. O. Atwater in the *Century Magazine* for May, June, July and September, 1887, and January and May, 1888. By the courtesy of the publishers with the consent of the author several of the subsequent tables have been copied from these articles.

---

*A Calorie is equivalent to about 1.53 foot-tons.

QUANTITY OF FOOD NEEDED DAILY IN ORDER THAT THE HUMAN BODY MAY KEEP ITSELF IN GOOD WORKING CONDITION.

Pages 35-55 and 58 are quoted, with slight alterations, from articles by Professor Atwater, in Vol. I of "The National Medical Dictionary" and "The Century" magazine, 1887 and 1888.

WAYS IN WHICH MATERIALS ARE USED IN THE BODY.

Protein forms tissue (muscle, tendon, etc., and fat) and serves as fuel,
Fats form fatty tissue (not muscle, etc.) and serve as fuel,
Carbohydrates are transformed into fat and serve as fuel,
} Yield energy in form of heat and muscular strength.

Alcohol does not form tissue, but does serve as fuel,
Tea and coffee (thein, caffein) do not form tissue, do not serve as fuel,
Extractives (meat-extract, beef-tea) do not form tissue, do not serve as fuel,
} Have various actions upon brain and nerves.

In being themselves burned to yield energy the nutrients protect each other from being consumed. The protein and fats of body-tissue are used like those of food. An important use of carbohydrates and fats is to protect protein (muscle, etc.) from consumption.

STANDARDS FOR DAILY DIETARIES FOR PEOPLE OF DIFFERENT CLASSES.

The demands of different people for nutrients in the daily food vary with age, sex, occupation, and other conditions, including especially the widely-differing characteristics of individuals. The standards in table IV., herewith, are intended to represent, roughly, the needs of average individuals of the classes named. Nos. 1, 3, 4, 5, and 6 are as proposed by Voit and his followers of the Munich school of physiologists, and are based upon observations of quantities actually consumed in a considerable number of cases. Nos. 7 and 8 are by Voit, and based both upon quantities consumed by individuals under experiment and upon observed dietaries of a much larger number of persons in Germany. Nos. 9, 10, and 11 are by Playfair, and are based mainly upon observations of actual dietaries in England. No. 2 is calculated by the writer from the data and results used in Nos. 1 and 3. In Nos. 12, 13, 14, and 15, by the writer, the data of Voit, Playfair, and other European observers are taken into account, but the conclusions are modified by the results of studies of a considerable number of dietaries in the United States, where people, wage-workers especially, are better fed, do more work, and receive higher wages.

Table IV.*

STANDARDS FOR DAILY DIETARIES FOR PEOPLE OF DIFFERENT CLASSES.

	Nutrients				Potential energy.
	Protein.	Fats.	Carbohydrates.	Total.	
	grams.	grams.	grams.	grams.	calories.
1 Children to 1½ years	28 (20 to 36)	37 (30 to 45)	75 (60 to 90)	140	765
2 Children 2 to 6 years	55 (36 to 70)	40 (35 to 48)	200 (100 to 250)	295	1420
3 Children 6 to 15 years	75 (70 to 80)	43 (37 to 50)	325 (250 to 400)	443	2040
4 Aged woman	80	50	260	390	1860
5 Aged man	100	68	350	518	2475
6 Woman at moderate work, German	92	44	400	536	2425
7 Man at moderate work, German	118	156	500	674	3055
8 Man at hard work, German	145	100	450	695	3370
9 Man with moderate exercise, English	119	51	531	701	3140
10 Active laborer, English	156	71	568	795	3630
11 Hard-worked laborer, English	185	71	568	824	3750
12 Women with light exercise, American	80	80	300	400	2300
13 Man with light exercise,† American	100	100	360	560	2815
14 Man at moderate work, American	125	125	450	700	3520
15 Man at hard work, American	150	150	500	800	4060

One pound avoirdupois=453.6 grams.   One ounce=28.3 grams.

Table V.

COMPARISON OF DIETARIES FOR CORPULENCE WITH ORDINARY DIETARIES.

Dietaries.	Nutrients				Potential energy in nutrients.
	Protein.	Fats.	Carbo-hydrates.	Total.	
	grams.	grams.	grams.	grams.	calories.
Banting system	171	8	75	254	1085
Ebstein system	102	85	47	234	1400
Playfair's standard for "subsistence diet"	57	14	341	412	1760
Playfair's standard for adults with moderate exercise	119	51	531	701	3140
Voit's standard for laboring-man at moderate work	118	56	500	674	3050
Writer's standard for man with light exercise	100	100	360	560	2815
Writer's standard for laboring-man at moderate work	125	125	450	700	3520
Poor sewing-girl, London; diet barely sufficient for subsistence.	53	33	316	402	1820
University professor, Germany, very little exercise	100	100	240	440	2225
Well-to-do family, Conn.; food actually eaten	128	177	466	771	4060
Mechanics and factory operatives, Mass.; food purchased	127	186	531	844	4430
College students, from northern and eastern U. S.; food actually eaten { I.	138	184	622	944	4825
II.	104	136	421	661	3415

* Several of the *Century* tables are omitted, but the same numbers are retained.
† Or woman with moderate work.

ACTUAL DIETARIES OF DIFFERENT CLASSES OF PEOPLE.

Table VI. gives the quantities of nutrients and potential energy in a number of observed dietaries. The figures for European dietaries are mostly by Voit and his followers in Germany, and by Playfair in England. The American figures are by the writer; those for the army and navy rations are based upon the United States Regulations, the rest upon observations of actual dietaries.*

TABLE VI.
NUTRIENTS AND POTENTIAL ENERGY IN DIETARIES OF DIFFERENT PEOPLE.

	Nutrients.				Potential energy of nutrients.
	Protein.	Fats.	Carbo-hydrates.	Total.	
	grams.	grams.	grams.	grams.	calories.
European and Japanese dietaries:					
1. Sewing-girl, London—wages 93 cts. (3s. 9d.) per week	53	33	316	402	1820
2. Factory-girl. Leipsic, Germany—wages $1.21 per week	52	53	301	406	1940
3. Weaver, England—time of scarcity	60	28	398	486	2138
4. Under-fed laborers, Lombardy, Italy—diet mostly vegetables	82	40	362	484	2192
5. Trappist monk in cloister; very little exercise—vegetable diet	68	11	469	548	2304
6. Students, Japan	97	16	438	551	2343
7. University professor, Munich, Germany; very little exercise	100	100	240	440	2324
8. Lawyer, Munich	80	125	222	427	2401
9. Physicians, Munich	131	95	327	553	2762
10. Painter, Leipsic, Germany	87	69	366	522	2500
11. Cabinetmaker, Leipsic, Germany	77	57	466	600	2757
12. "Fully-fed " tailors, England	181	39	525	625	3053
13. "Well-paid " mechanic, Munich, Germany	151	54	479	684	3065
14. Carpenter, Munich, Germany	131	68	494	693	3194
15. "Hard-worked" weaver, England	151	43	622	816	3569
16. Blacksmith, England	176	71	667	914	4117
17. Miners at very severe work, Germany	133	113	634	880	4195
18. Brickmakers (Italians at contract-work), Munich	187	117	675	959	4541
19. Brewery laborer, Munich; very severe work—exceptional diet	223	113	909	1245	5692
20. German soldiers, peace-footing	114	39	480	633	2798
21. German soldiers, war-footing	134	58	489	681	3093
22. German soldiers, Franco-German War; extraordinary ration	157	285	331	773	4652
United States and Canadian dietaries:					
23. French Canadians, working people, in Canada	109	109	527	745	3622
24. French Canadians, factory operatives, mechanics, etc., in Mass.	118	204	549	871	4632
25. Other factory operatives, mechanics, etc., Mass.	127	186	531	844	4428
26. Glassblowers, East Cambridge	95	132	481	708	3500
27. Factory operatives, dressmakers, clerks, etc., b'rd'g-house, Mass.	114	150	522	786	4002
28a. Well-to-do private family { Food purchased	129	183	467	779	4146
28b. { Food eaten	128	177	406	711	4062
29a. College students from northern and { Food purchased	161	204	680	1045	5345
29b. eastern States; boarding-club, two { Food eaten	138	164	622	944	4827
30a. dietaries of the same club { Food purchased	115	163	460	738	3674
30b. { Food eaten	104	136	421	661	3417
31. College foot-ball team, food eaten	181	292	557	1030	5742
32. Machinist, Boston, Mass.	182	254	617	1053	5638
33. Brickmakers, Middletown, Conn.	222	263	758	1243	6464
34. Teamsters, marble-workers, etc., with hard work; Boston, Mass.	254	363	826	1443	7804
35. Brickmakers, Mass.	180	365	1150	1695	8848
36. U. S. army ration	120	161	454	735	3851
37. U. S. navy ration	143	184	520	847	4098

*Comparison of the data of tables IV., V., and VI. implies that the food of people in this country is apt to be considerably in excess of the demands for nourishment. The excess is made up largely of meats, especially of the fatter kinds, and of sweetmeats, including sugar.

Standards for Daily Dietaries and Actual Dietaries of People of Different Classes.

WEIGHTS OF NUTRIENTS AND CALORIES OF ENERGY IN NUTRIENTS IN FOOD PER DAY.

PROTEIN. Lean of meat, casein of milk, white of egg, gluten of wheat, etc.	FATS. Fatty and Oily Matters.	CARBOHYDRATES. Sugar, Starch, etc.	POTENTIAL ENERGY Fuel Value.

NUTRIENTS, GRAMS — 200, 400, 600, 800, 1000, 1200, 1400, 1600
POTENTIAL ENERGY CALORIES — 1000, 2000, 3000, 4000, 5000, 6000, 7000, 8000

DIETARY STANDARDS.
- Subsistence diet (Playfair)
- Man at moderate work (Voit)
- Man at hard work (Voit)
- Man with moderate exercise (Playfair)
- Man with light exercise (Atwater)
- Man at moderate work (Atwater)
- Man at hard work (Atwater)

ACTUAL DIETARIES.
- Under-fed laborers, Lombardy, Italy
- Students, Japan
- Lawyer, Munich, Germany
- Physician, Munich, Germany
- Well-paid Mechanic, Munich, Germany
- Carpenter, Munich, Germany
- Well-fed Blacksmith, England
- German soldiers, peace footing
- German soldiers, war footing
- French-Canadian families, Canada
- Mechanics and factory operatives, Mass.
- Well-to-do family, Conn.
- College students, Northern and Eastern States
- Machinist, Boston, Mass.
- Hard worked teamsters, etc., Boston, Mass.
- U. S. Army Ration
- U. S. Navy Ration

W. O. ATWATER.

National Medical Dictionary.
Reprinted by consent.

COMPOSITION OF SOME FOOD MATERIALS FROM WHICH THIS SUPPLY IS DERIVED.

By Professor W. O. Atwater.

COMPOSITION, DIGESTIBILITY, AND POTENTIAL ENERGY (FUEL-VALUE) OF FOOD-MATERIALS, FUNCTIONS OF NUTRIENTS, DIETARY STANDARD AND ACTUAL DIETARIES.

The following tables and explanatory statements are an attempt to epitomize some results of late research, a considerable portion of which have not yet become current in treatises in English.* Those regarding digestibility, potential energy, and functions in nutrients are based upon experimental inquiry in Europe, especially in Germany. The data employed for the Tables of dietary standards and actual dietaries are mainly European, but include a considerable number of results of observations in the United States. The figures for the composition of food-materials are taken from American analyses. The larger number of the latter were executed in connection with investigations in behalf of the Smithsonian Institution and the United States Fish Commission, but still await detailed publication.

INGREDIENTS OF FOOD-MATERIALS—NUTRIENTS AND NON-NUTRIENTS.

Our ordinary food-materials such as meat, fish, eggs, potatoes, wheat, etc. consists of—

Refuse, as the bones of meat and fish, shells of eggs, skins of potatoes, and bran of wheat.

Edible portion, as the flesh of meat and fish, white and yolk of eggs, wheat flour.

The edible substance consists of—
    Water,
    Nutritive ingredients or nutrients.
The principal kinds of nutrient are—
    1. *Protein;*
    2. *Fats;*
    3. *Carbohydrates;*
    4. *Mineral matters.*

---

*See articles on the Chemistry of Foods and Nutrition in the *Century Magazine* for 1887 and 1888.

40   THE SCIENCE OF NUTRITION.

Water, refuse, and the salt of salted meat and fish are called non-nutrients. The water contained in foods and beverages has the same composition and properties as other water; it is, of course, indispensable for nourishment, but is not a nutrient in the sense in which the term is here used. In comparing the values of different food-materials for nourishment, the refuse and water are left out of account.

CLASSES OF NUTRIENTS.

The following are familiar examples of compounds of each of the four principal classes of nutrients:

Protein.
- *a.* Albuminoids: *e. g.* albumen (white of eggs); casein (curd) of milk; myosin, the basis of muscle (lean meat); gluten of wheat, etc.
- *b.* Gelatinoids: *e. g.* collagen of tendons; ossein of bones, which yield gelatin or glue.
  (Meats and fish contain very small quantities of so-called "extractives." They include kreatin and allied compounds, and are the chief ingredients of beef-tea and meat-extract. They contain nitrogen, and hence are commonly classed with protein.

Fats. . . *e. g.* fat of meat; fat (butter) of milk; olive oil; oil of corn, wheat, etc.

Carbo-hydrates. *e. g.* sugar, starch, cellulose (woody fiber).

Mineral matters. *e. g.* calcium phosphate or phosphate of lime; sodium chloride (common salt).

It is to be especially noted that the protein compounds contain nitrogen, while the fat and carbohydrates have none. The albuminoids and gelatinoids are frequently classed together as proteids. The term "proteids" is also used to include all the nitrogenous ingredients—*i. e.* synonymous with protein. The average composition of these compounds is about as follows:

	Protein.	Fats.	Carbohydrates.
Carbon,	53 per cent.	76.5 per cent.	44 per cent.
Hydrogen,	7 "	12.0 "	6 "
Oxygen,	24 "	11.5 "	50 "
Nitrogen,	16 "	None.	None.
	100 per cent.	100.0 per cent.	100 per cent.

POTENTIAL ENERGY OF FOOD.

In being consumed in the body as fuel to furnish heat and muscular energy the nutrients appear to replace one another in proportion to their potential energy, which is accordingly taken as a measure of their fuel-value. The energy is estimated in Calories. The Calorie is the heat which would raise a kilogram of water one degree centigrade (or one pound of water about four degrees Fahrenheit). A foot-ton is the energy (power) which would lift one ton one foot. One Calorie corresponds to 1.53 foot-tons. A gram of protein or a gram of carbohydrates is assumed to yield 4.1, and a gram of fats 9.3 Calories. A given weight of fats is thus taken to be equivalent in fuel-value, on the average, to a little over twice the same weight of protein or carbohydrates. The figures for potential energy in Table I. are calculated for each food-material by multiplying the number of grains of protein and of carbohydrates in one pound (1 lb. equals 453.6 grams) by 4.1, and the number of grams of fat by 9.3, and taking the sum of these three products as the number of Calories of potential energy in a pound of the material.

COMPOSITION OF FOOD-MATERIALS.

Different specimens of the same kind of food-material differ widely in composition. The figures in Table I., herewith represent the averages of analyses, of which those of fruits and beverages are European and the rest American.

TABLE I.

PERCENTAGES OF NUTRIENTS (NUTRITIVE INGREDIENTS), WATER, ETC., AND ESTIMATED POTENTIAL ENERGY (FUEL-VALUE) IN SPECIMENS OF FOOD-MATERIALS.

FOOD-MATERIALS.	REFUSE: bones, skin, shells, etc.	Edible Portion.						Calories of potential energy in one pound of each material.
		Water.	Nutrients.					
			Total.	Protein.	Fats.	Carbo-hydrates.	Mineral matters.	
Animal foods as purchased, including edible portion and refuse:	per cent.	per cent.	per cent.	per cent.	per cent.	per cent.	per cent.	
Beef, side[1]	19.7	44.0	36.3	13.8	21.7	. .	0.8	1170
Beef, round[1]	10.0	60.0	30.0	20.7	8.1	. .	1.2	725
Beef, neck[1]	19.9	49.6	30.5	15.4	14.3	. .	0.8	890
Beef, sirloin[1]	25.0	45.0	30.0	15.0	14.3	. .	0.7	885
Beef, flank[1]	11.7	24.2	64.1	10.6	52.9	. .	0.6	2430
Mutton, side[1] . . .	20.0	42.9	37.1	13.2	23.2	. .	0.7	1225
Mutton, leg[1]	18.4	50.4	31.2	15.0	15.5	. .	0.7	935
Mutton, shoulder[1] . .	16.8	48.7	34.5	15.0	18.7	. .	0.8	1070
Mutton, loin (chops)[1] .	16.3	41.3	42.4	12.5	29.3	. .	0.6	1470
Smoked ham, . . .	14.0	36.3	49.7	14.6	34.2	. .	0.9	1715
Pork, very fat . . .	10.4	9.5	80.1	2.8	76.5	. .	0.8	3280
Chicken[2]	41.6	42.2	16.2	14.2	1.2	. .	0.8	315
Turkey	35.4	42.8	21.8	15.4	5.6	. .	0.8	525
Flounder, whole . .	66.8	27.2	6.0	5.2	0.3	. .	0.5	110
Haddock, dressed . .	51.0	40.0	9.0	8.2	0.2	. .	0.6	160
Bluefish, dressed . .	48.6	40.3	11.1	9.8	0.6	. .	0.7	210
Brook trout, whole .	48.1	40.4	11.5	9.8	1.1	. .	0.6	230
Codfish, dressed . .	29.9	58.5	11.6	10.6	0.2	. .	0.8	205
Whitefish, whole . .	53.5	32.5	14.0	10.3	3.0	. .	0.7	320
Shad, whole	50.1	35.2	14.7	9.2	4.8	. .	0.7	375
Turbot, whole . . .	47.7	37.8	15.0	6.8	7.5	. .	0.7	445
Mackerel, fat, whole, .	33.8	42.4	23.8	12.1	10.7	. .	1.0	675
Mackerel, lean, whole	38.3	48.5	13.2	11.2	1.4	. .	0.6	265
Mackerel, average, whole	44.6	40.4	15.0	10.0	4.3	. .	0.7	365
Halibut, dressed . .	17.7	61.9	20.4	15.1	4.4	. .	0.9	465
Salmon, whole . . .	35.3	40.6	24.1	14.3	8.8	. .	1.0	635
Eel	36.0	33.8	30.2	8.6	21.0	. .	0.6	1045
Salt codfish	42.1	40.3	17.6	16.0	0.4	. .	1.2	315
Smoked herring . . .	50.9	19.2	29.9	20.2	8.8	. .	0.9	745
Salt mackerel . . .	40.4	28.1	31.5	14.7	15.1	. .	1.7	910
Canned salmon . . .	4.9	59.3	35.8	19.3	15.3	. .	1.2	1005
Canned sardines . .	5.0	53.6	41.4	24.0	12.1	. .	5.3	955
Lobsters	62.1	31.0	6.9	5.5	0.7	0.1	0.6	135
Oysters in shell . . .	82.3	15.4	2.3	1.1	0.2	0.6	0.4	40
Hens' eggs	13.7	63.1	23.2	11.8	10.2	0.4	0.8	655
Animal foods, edible portion:								
Beef, side[1]	. .	54.7	45.3	17.2	27.1	. .	1.0	1465
Beef, round[1]	. .	66.7	33.3	23.0	9.0	. .	1.3	805
Beef, sirloin[1]	. .	60.0	40.0	20.0	19.0	. .	1.0	1175
Mutton, side[1]	. .	45.9	54.1	14.7	38.7	. .	0.7	1905
Mutton, leg[1]	. .	61.8	38.2	18.3	19.0	. .	0.9	1140
Mutton, loin (chops)[1] .	. .	49.3	50.7	15.0	35.0	. .	0.7	1755

[1] From well-fattened animals. [2] Rather lean.

TABLE I.—Concluded.

FOOD-MATERIALS	REFUSE: bones, skin, shells, etc.	EDIBLE PORTION.						Calories of potential energy in one pound of each material.
		Water.	Nutrients.					
			Total.	Protein.	Fats.	Carbo-hydrates.	Mineral matters.	
	per cent.	per cent.	per cent.	per cent.	per cent.	per cent.	per cent.	
Animal foods, edible portion:								
Flounder	..	84.2	15.8	13.8	0.7	..	1.3	285
Codfish	..	82.6	17.4	15.8	0.4	..	1.2	310
Mackerel, fat	..	64.0	36.0	18.2	16.3	..	1.5	1025
Mackerel, lean	..	78.7	21.3	18.1	2.2	..	1.0	430
Mackerel, average	..	71.6	28.4	18.8	8.2	..	1.4	695
Salmon	..	63.6	36.4	21.6	13.4	..	1.4	965
Oysters, fat	..	81.7	18.3	8.0	1.7	6.7	1.9	345
Oysters, lean	..	90.9	9.1	4.2	0.6	1.8	2.5	135
Oysters, average	..	87.1	12.9	6.0	1.2	3.7	2.0	280
Hens' eggs	..	73.1	26.9	13.7	11.7	0.4	1.0	760
Cows' milk	..	87.4	12.6	3.4	3.7	4.8	0.7	310
Cows' milk	..	90.7	9.3	3.1	0.7	4.8	0.7	175
Cheese, whole milk	..	31.2	68.8	27.1	35.5	2.3	3.9	2045
Cheese, skimmed milk	..	41.3	58.7	38.4	6.8	8.9	4.5	1165
Butter	..	10.0	90.0	1.0	85.0	0.5	3.5	3615
Oleomargarine	..	10.0	90.0	0.6	84.5	0.4	4.5	3585
Lard	..	1.0	99.0	..	99.0	..	..	4180
Beverages:								
Wheat bread	..	32.7	67.3	8.9	1.9	55.5	1.0	1280
Wheat flour	..	11.6	88.4	11.1	1.1	75.6	0.6	1660
Graham flour	..	13.0	87.0	11.7	1.7	71.8	1.8	1625
Rye flour	..	13.1	86.9	6.7	0.7	78.7	0.7	1620
Buckwheat flour	..	13.5	86.5	6.5	1.3	77.6	1.1	1620
Beans	..	13.7	86.3	23.2	2.1	57.4	3.6	1585
Oatmeal	..	7.7	92.3	15.1	7.1	68.1	2.0	1845
Corn (maize) meal	..	14.5	85.5	9.1	3.8	71.0	1.6	1650
Rice	..	12.4	87.6	7.4	0.4	79.4	0.4	1630
Sugar	..	2.2	97.8	0.3	..	96.7	0.8	1800
Potatoes[1]	10.0	68.0	22.0	1.8	0.2	19.1	0.9	395
Potatoes	..	75.5	24.5	2.0	0.2	21.3	1.0	440
Sweet potatoes	..	75.8	24.2	1.5	0.4	21.1	1.2	435
Turnips	..	91.2	8.8	1.0	0.2	6.9	0.7	155
Carrots	..	87.9	12.1	1.0	0.2	10.1	0.8	215
Cabbage	..	90.0	10.0	1.9	0.2	6.2	1.2	170
Melons	..	95.2	4.8	1.1	0.6	2.5	0.6	90
Apples	..	84.8	15.2	0.4	..	14.3	0.5	275
Pears	..	83.0	17.0	0.4	..	16.3	0.3	310
Bananas	..	73.1	26.9	1.9	0.6	23.3	1.1	495
Vegetable foods:						Alcohol.		
Lager beer	..	90.3	..	0.4	..	2.0 / 5.8	0.2	..
Porter and ale	..	88.1	..	0.6	..	5.1 / 6.8	0.4	..
Rhine wine, white	..	86.3	..	..	..	9.3 / 2.3	0.2	..
Rhine wine, red	..	86.9	..	..	..	8.1 / 3.0	0.3	..
French wine, claret	..	88.3	..	..	..	8.0 / 2.3	0.2	..
Sherry wine	..	79.5	..	..	..	17.0 / 3.2	0.3	..

[1] As purchased, including refuse, skin, etc.

DIGESTIBILITY OF FOODS.

Table II. epitomizes the results of some sixty experiments, mostly with men, but a few with children, in which the proportions of the ingredients of food-materials actually digested have have been found by comparison of amounts and composition of the food eaten with those of the undigested excreta. Table III. is computed by applying the data obtained by these experiments to some of those for the composition of food-materials in Table I.

TABLE II.

DIGESTIBILITY OF NUTRIENTS OF FOOD-MATERIALS.

In the Food-Materials Below.	Of the total amounts of protein, fats, and carbohydrates, the following percentages were digested.		
	Protein.	Fats.	Carbo-hydrates.
Meats and fish	Practically all.	79 to 92	
Eggs	"	96	
Milk	88 to 100	93 to 98	?
Butter	. . . .	98	
Oleomargarine	. . . .	96	
Wheat bread	81 to 100	?	99
Corn (maize) meal	89	?	97
Rice	84	?	99
Pease	86	?	96
Potatoes	74	?	92
Beets	72	?	82

TABLE III.

PROPORTIONS OF NUTRIENTS DIGESTED AND NOT DIGESTED FROM FOOD-MATERIALS BY HEALTHY MEN.

	PROTEIN.			FATS.			CARBOHYDRATES.			Mineral matters.	Water.
	Digestible.	Undigestible.	Total.	Digestible.	Undigestible.	Total.	Digestible.	Undigestible.	Total.		
	per ct.	per ct.	per ct.	per ct.	per ct.	per ct.	per ct.	per ct.	per ct.	per ct.	per ct.
Beef, round	23.0	0.0	23.0	8.1	0.9	9.0	0.0	0.0	0.0	1.3	66.7
Beef, sirloin	20.0	0.0	20.0	17.1	1.9	19.0	0.0	0.0	0.0	1.0	60.0
Pork, very fat	3.0	0.0	3.0	74.5	6.0	80.5	..	..	..	6.5	10.0
Haddock	17.1	0.0	17.1	0.3	..	0.3	0.0	0.0	0.0	1.2	81.4
Mackerel	18.8	0.0	18.8	7.4	0.8	8.2	0.0	0.0	0.0	1.4	71.6
Hens' eggs	13.4	0.0	13.4	9.4	2.4	11.8	0.7	0.0	0.7	1.0	73.1
Cows' milk	3.4	0.0	3.4	3.6	0.1	3.7	4.8	0.0	4.8	0.7	87.4
Cheese, whole milk	27.1	0.0	27.1	34.6	0.9	35.5	2.3	0.0	2.3	3.9	31.2
Butter	1.0	..	1.0	85.8	1.7	87.5	0.5	..	0.5	2.0	9.0
Oleomargarine	0.4	..	0.4	83.9	3.3	87.2	0.0	..	0.0	2.1	10.3
Sugar	0.3	..	0.3	..	..	..	96.7	0.0	96.7	0.8	2.2
Wheat flour, very fine	7.6	1.3	8.9	1.0	..	1.0	74.4	0.8	75.2	0.3	14.6
Wheat flour, medium	9.5	2.1	11.6	0.8	..	0.8	70.4	1.8	72.2	0.4	15.0
Wheat flour, coarse, whole wh.	8.2	2.7	10.9	1.8	..	1.8	66.4	5.3	71.7	1.2	14.4
Wheat, bread average	7.7	1.2	8.9	1.9	..	1.9	54.9	0.6	55.5	1.0	32.7
Black bread	4.5	1.6	6.1	..	..	..	43.3	5.3	48.6	1.5	43.8
Pease	19.7	3.2	22.9	1.8	..	1.8	55.7	2.1	57.8	2.5	15.0
Corn (maize) meal	7.9	1.2	9.1	3.8	..	3.8	68.7	2.3	71.0	1.6	14.5
Rice	6.2	1.2	7.4	0.4	..	0.4	78.7	0.7	79.4	0.4	12.4
Potatoes	1.5	0.5	2.0	0.2	..	0.2	19.7	1.6	21.3	1.0	75.5
Turnips	0.7	0.3	1.0	0.2	..	0.2	5.6	1.3	6.9	0.7	91.2

USES OF FOOD IN THE BODY.

Food supplies the wants of the body in several ways. It either—
1. Is used to form the tissues and fluids of the body;
2. Is used to repair the wastes of tissue;
3. Is stored in the body for future consumption;
4. Is consumed as fuel, its potential energy being transformed into heat or muscular energy or other forms of energy required by the body; or
5. In being consumed protects tissue or other food from consumption.

One can hardly present the elements of this case in too many aspects and as the purpose of this treatise is to promote further work on the same line, various tables will be given at the risk of repetition, each one dealing with the matter under a different aspect.

In dealing with the Pecuniary Economy of food in No. V of the Century Articles, Prof. Atwater makes the following quotation and then proceeds to show what true economy may be:

"No one can say that I do not give my family the best of flour, the finest sugar, the very best quality of meat."

The above is the boast of a coal laborer earning seven dollars a week. It illustrates a phenomenon which I would commend to the consideration of either psychologists or students of social science, or both. I refer to the conceit, let us call it, that there is some mysterious virtue in those kinds of foods that have the most delicate appearance and flavor and the highest price; that whatever else one has or does not have he must, if possible, have this sort of food; and that to economize by using anything inferior would be a sacrifice of both dignity and principle.

The quotation, from a description of the life of factory operatives in New England, in an article by Mr. Lee Meriwether, in "Harper's Magazine" for April, 1887, illustrates what I mean.

The cheapest food is that which supplies the most nutriment for the least money. The most economical food is that which is cheapest and best adapted to the wants of the user. But the maxim that "the best is the cheapest" does not apply to food. The best food, in the sense of that which has the finest appearance and flavor and is sold at the highest price, is not generally the cheapest nor the most economical, nor is it always the most healthful. The coal laborer who made it so much an article of faith to give his family "the best of flour, the finest sugar, the very best quality of meat;" who, as Mr. Meriwether tells us, at a time when excellent butter was selling at 25 cents a pound paid 29 cents for an extra quality; who spent $156 a year for the nicest cuts of meat, which his wife had to cook before six in the morning or after half-past six at night because she worked all day in the factory; who spent only $108 for clothing for his family of nine, and only $72 a year for rent in a crowded tenement-house where they slept in rooms without windows or closets; who indulged in this extravagance in food when much cheaper meat and in all probability much less of it, cheaper butter, cheaper flour, and other less costly materials such as come regularly upon the table of many a man of wealth would have been just as wholesome, just as nutritious, and in every way just as good save in its gratification to pride and palate,—this man was innocently committing an immense economical and hygienic blunder. He was doing this because, like the very large class of people of whom he is a type, he was laboring under this conceit of which I speak.

THE SCIENCE OF NUTRITION.

One great difficulty here is the lack of information. Even those who wish and try to economize in the purchase and use of food very often do not understand how. They consult carefully the prices they pay, but have in general very vague ideas about the nutritive values. It is an interesting fact that although the cost of food is the principal item of the living expenses of the large majority of people,—of all, indeed, but a few of the especially well-to-do,*—and although the health and strength of all are so intimately dependent upon their diet, yet even the most intelligent know less of the actual uses and value of their food for fulfilling its purposes than of those of almost any other of the staple necessities of life.

PERCENTAGE OF INCOME EXPENDED FOR SUBSISTENCE.

Families of	Annual Income.	Per cent. expended for food.
GERMANY.		
Workingmen,	$225 to $300	62
Intermediate class, "*Mittelstand*,"	450 to 600	55
In easy circumstances, "*Wohlstand*,"	750 to 1100	50
GREAT BRITAIN.		
Workingmen,	500	51
MASSACHUSETTS.		
Workingmen,	350 to 400	64
"	450 to 600	63
"	600 to 750	60
"	750 to 1200	56
"	above 1200	51

RATIOS OF NUTRITIVE VALUES TO COST.

The large majority of the families in this country have, I understand, not over $500 a year to live upon. More than half of this goes, and must go, for food. Rent, clothing, the cost of preparing the food for the table, and all other expenses must be provided from the rest. Perhaps these statements apply less accurately to farmers, but of wage-workers in towns statisticians tell me they are correct.

To the man with an income of $5,000 a year, it may seem to make little differ-

*In his Report of the Bureau of Statistics of Labor of Massachusetts for 1884, Mr. Carroll D. Wright summarizes the results of investigations into the cost of living of people with different incomes, especially of workingmen's families, in Massachusetts and in Great Britain, and quotes similar results obtained by Dr. Engel in Germany. Dividing expenses into those for subsistence (food), clothing, rent, fuel, and sundries, the percentage of the whole income expended for subsistence averages as in the tabular statement herewith. As incomes increase the relative percentage of outlay for food becomes less and that for "sundries" greater. In the Massachusetts and Great Britain figures (I do not know how it is with the German, but presume that the case is the same,) no outlay for intoxicating liquors is included in the allowance for subsistence.

ence whether he pays 20 cents or $2 a pound for the protein of his food; but to the one who can earn only $500 or less a year for the support of his family, the difference is an important one. His wife goes to the dry goods store to buy a dress for her daughter, and hesitates between a piece of cloth at 40 cents a yard that would please her better and one at 35 that is not so pretty but just as durable, and is very apt to take the cheaper one because she feels that she must. She does not fall into the error of getting more cloth than is needed and using part of the excess for lining and throwing the rest away, nor, if she is wise, does she try to economize by getting poor trimmings and cheap thread. But when she goes to the grocer or to the butcher or to the fish market for food to build up her children's bodies and give her husband and herself strength to work, she often pays one or two dollars a pound for protein to make muscle when she might obtain it in forms equally wholesome and nutritious for from 15 to 50 cents. The food she buys is apt to supply some of the nutrients in excessive amount as well as at needlessly high cost, while it furnishes others in insufficient quantity or in unfitting forms and in uneconomical ways; and only too often a part of it finds its way into the drain or the garbage barrel instead of being utilized for nourishment.

Of course the good wife and mother does not understand about protein and potential energy and the connection between the nutritive value of food and the price she pays for it, and doubtless she never will. But if the knowledge is obtained and put in print, and diffused among those who have the time and training to get hold of it, the main facts will gradually work their way to the masses, who most need its benefit.

A subject that has received but little attention in this country, though it is one of the many special problems that are being carefully considered by students of social economy in Europe, is the relation of the nutritive value of food to its cost. We purchase our food by gross weight or measure. Part of it consists of nutritive substances, the rest is made up of water and various materials which serve only as ballast. In comparing different food-materials with respect to their cheapness or dearness we are apt to judge them by the prices per pound, quart, or bushel, without much regard to the amounts or kinds of actual nutrients which they contain. Of the different food-materials which the market affords and which are palatable, nutritious, and otherwise fit for nourishment, what ones are pecuniarily the most economical?

In a series of studies, undertaken at the instance of the Smithsonian Institution, I have had occasion to examine into some of these problems. A few of the results of the inquiry are summarized in Diagrams VI. and VII.

There are various ways of comparing food-materials with respect to the relative cheapness or dearness of their nutritive ingredients. The best, perhaps, consists in simply comparing the quantities of nutrients obtained for a given sum, 25 cents for instance, in the food when purchased at market prices. Diagram VI. gives a series of such comparisons. They are based upon the analyses of materials, obtained mostly in markets in New York City and in Middletown, Conn., and upon the retail prices paid for them. Along with the quantities of nutrients which 25 cents will buy are shown the quantities estimated to be appropriate for a day's diet for an ordinary man doing a moderate amount of muscular labor. Two such standards are given—one proposed by Professor Voit in Germany, and based mainly upon experiments and observations in that country; the other proposed by myself. The diagram shows the quantities of different food-materials which one would get for a quarter of a dollar; the quantities of protein and fats and carbohydrates contained in them; and how these amounts of nutrients compare with what an average man, engaged in moderately hard muscular work, might be expected to need to maintain his body in vigorous condition and supply strength for the work he has to do. Another way of comparing the nutritive value of the food-materials with the cost is by the quantities of potential energy they contain. Diagram VII. shows the estimated quantities of energy in the nutritive ingredients of the materials in Diagram VI.,—that is, the amount which 25 cents would pay for. Still another method of comparing the actual expensiveness of different foods at the prices at which people buy them consists in comparing the cost of the same nutrient in different food-materials. Of the different nutrients, protein is physiologically the most important, as it is pecuniarily the most expensive. For these reasons the cost of protein in different food-materials may be used as a means of comparing their relative cheapness or dearness, as is done in Diagram VII. The figures represent the ordinary prices per pound and the corresponding costs of protein, due allowance being made for the carbohydrates and fats, the estimated costs of which are, for the sake of brevity, omitted from the table.*

* As explained in previous articles, the actually nutritive ingredients of food may be divided into four classes: Protein, Fats, Carbohydrates, and Mineral Matters. Leaving water out of account, lean meat, white of eggs, casein (curd) of milk, and gluten of wheat consist mainly of protein compounds. Butter and lard are mostly fats. Sugar and starch are carbohydrates. The nutrients of meat, fish, and other animal foods consist mainly of protein and fats; those of the vegetable foods are largely carbohydrates.

In serving as nutriment, the protein compounds which contain nitrogen form the basis of blood, muscle, tendon, etc., and are transformed into fat, and also serve as fuel to supply the body with heat and muscular strength. The fats of the food are stored as fat in the body and serve as fuel. The carbohydrates are transformed into fats and serve as fuel. The potential energy in Calories (calorie is the equivalent of heat which would warm about four pounds of water one degree Fahrenheit) is taken as the measure of the fuel-value of the food. One part by weight of fat is equivalent, in this respect, to about two parts of either protein or carbohydrates. The demands of different people for nourishment vary with age, sex, occupation, and other conditions of life. Health and pecuniary economy alike require that the diet should contain nutrients proportionate to the wants of the user. Of course the difference in the composition of different specimens of the same kind of food-material, and

EXPENSIVE VS. ECONOMICAL FOODS.

Taking the diagrams and tabular statements together, the first thing that strikes one is the cheapness of the vegetable as compared with the animal foods. Note, for instance, Diagram VI. and the accompanying figures, which show how much actually nutritive material one may have for 25 cents in different foods at ordinary prices. The quarter of a dollar invested in flour, meal, or potatoes brings several times the quantity of nutrients that it does if spent for meats, fish, or milk. But it is to be remembered that the animal foods contain more of the protein and fats, which are the most valuable food constituents, while the excess of material obtained in the vegetable foods consists mainly or entirely of sugar, starch, and other carbohydrates, which, though very important for nourishment, are far less valuable, weight for weight, than the protein and fats. Furthermore, the protein of the animal foods is more easily and completely digestible than that of the vegetable foods.

The greater expensiveness of animal foods is brought out with even greater clearness in Diagram VII. and in the accompanying figures. The quantities of potential energy in the nutritive material obtained for 25 cents range, in the animal foods, from 160 Calories, in the salmon at a dollar a pound, to 6,800, in salt pork at 13 cents a pound; while in the vegetable foods in the tables the range is from about 500, in rice at 8 cents a pound, to 1,200, in corn meal at 2 cents a pound. The standards for the diets of an ordinary workingman call for from 3,000 to 3,600 Calories in one day's food.

Estimating the expensiveness by the cost of the protein, we find this to range from 8 to 34 cents a pound in the vegetable, and from 18 cents to a little over one dollar in ordinary animal foods—meats, fish, milk, eggs, etc.—while in some it is much higher, thus showing the greater expensiveness of animal foods in another way. The reason for this higher cost is, of course, simple enough. Animal foods are made from vegetable, and by a more or less expensive process. The manufacture of beef or milk from grass and grain involves considerable outlay for labor and incidental expenses, and the product is, of course, much less in quantity than the raw material.

If the reader is interested in such statistics he will find considerable food for reflection in the diagrams and figures. He will observe that among animal foods

---

in the nutritive effect of the same substance with different persons, is such that these calculations are not correct for every case. Furthermore, there are other things besides the proportions of nutrients that affect the nutritive action of food. This topic I hope to discuss later. Meanwhile it will suffice to say that for the staple food-materials these calculations are probably close approximations to the real nutritive values as compared with the costs The methods by which they are made are too complex to be explained here, but may be found in an article on "Food Consumption" in the Report of the Massachusetts Bureau of Statistics of Labor for 1886, p. 251.

DIAGRAM VI.—COMPARATIVE EXPENSIVENESS OF FOODS.

Amounts of Actual Nutrients (Nutritive Ingredients) Obtained for Twenty-five Cents in Different Food-Materials at Ordinary Prices, with Amounts Appropriate for a Day's Ration.

Food-Materials.	*Assumed prices per pound* Cts.	*Food-Materials.* Lbs.
Beef, sirloin	.25	1.00
Beef, sirloin, at lower price	.20	1.25
Beef, round	.16	1.56
Beef, neck	.08	3.13
Mutton, leg	.22	1.14
Smoked Ham	.14	1.79
Salt Pork, very fat	.12	2.08
Salmon, early in season	1.00	.25
Salmon, at lower price	.30	.83
Mackerel	.10	2.50
Codfish	.08	3.13
Salt Mackerel	.12½	2.00
Salt Codfish	.07	3.57
Oysters, at 40 cents per quart	.20	1.25
Hens' eggs, at 30 cents per dozen	.21⅘	1.15
Milk, at 7 cents per quart	.03½	7.14
Cheese, whole milk	.15	1.67
Cheese, skimmed milk	.08	3.13
Butter	.30	.83
Oleomargarine	.15	1.67
Sugar	.07⅐	3.50
Wheat flour	.03	8.33
Wheat bread	.07½	3.33
Corn (maize) meal	.03	8.33
Oatmeal	.05	5.00
Rice	.08	3.13
Beans	.05	5.00
Potatoes, at 75 cents per bushel	.01¼	20.00
Standards for daily diet for laboring man at moderate work — Voit's, German		
Standards for daily diet for laboring man at moderate work — Writer's, American		

Quantities obtained for 25 cents.

Nutrients in the Food-Materials.
Quantities in pounds and hundredths of a pound indicated by shaded bands.
PROTEIN. "Lean" of meat, gluten of wheat, etc.
FATS. Fatty and oily substances.
CARBOHYDRATES. Sugar, starch, etc.
.50
1 lb.
1.50
2 lbs.
2.50
3 lbs.
3.50
4 lbs.

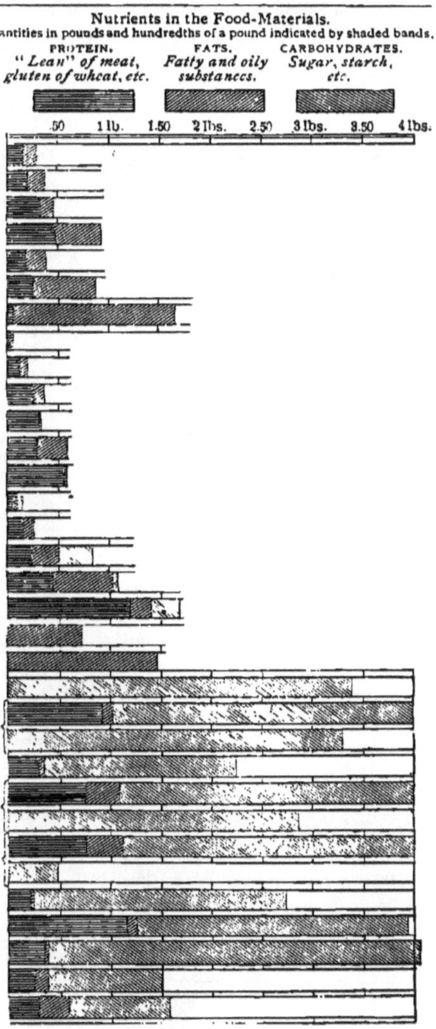

W. O. ATWATER.

Century Magazine.
Reprinted by consent.

DIAGRAM VII.—COMPARATIVE EXPENSIVENESS OF FOODS.

Costs of a Pound of Protein and Amounts of Potential Energy Obtained for Twenty-five Cents in Different Food-Materials at Current Market Prices.

Kinds of Food-Materials.	Assumed prices of Food-materials per pound in cents.	The estimated cost (in cents) of one pound of protein in each Food-material, when the latter is bought at the market prices assumed, is expressed by the lengths of the light parallel lines, thus: ▭ The estimated number of calories of potential energy in the nutrients (actually nutritive ingredients) contained in the quantity of each Food-material which 25 cents would pay for if the material were bought at the market prices assumed, is expressed by dark lines, thus: ▬
Beef, sirloin	.25	106 cents... 870 calories
Beef, sirloin, at lower price	.20	85 cents... 1114 calories
Beef, round	.16	63 cents... 1145 calories
Beef, neck	.08	33 cents... 2793 calories
Mutton, leg	.22	91 cents... 1076 calories
Smoked Ham	.14	40 cents... 3060 calories
Salt Pork, very fat	.12	25 cents... 6825 calories
Salmon, early in season	1.00	511 cents... 128 calories
Salmon, at lower price	.30	153 cents... 510 calories
Mackerel	.10	79 cents... 929 calories
Codfish	.08	75 cents... 656 calories
Salt mackerel	.12½	72 cents... 1807 calories
Salt codfish	.07	43 cents... 1103 calories
Oysters, at 40 cents per quart	.20	258 cents... 326 calories
Hens' eggs, at 30 cents per dozen	.21⅘	121 cents... 765 calories
Milk, at 7 cents per quart	.03½	53 cents... 2178 calories
Cheese, whole milk	.15	11 cents... 3403 calories
Cheese, skimmed milk	.08	18 cents... 3642 calories
Butter	.30	no protein. 3082 calories
Oleomargarine	.15	no protein. 6164 calories
Sugar	.07	no protein. 6292 calories
Wheat flour	.03	11 cents... 13782 calories
Wheat bread	.07½	35 cents... 4255 calories
Corn (maize) meal	.03	12 cents... 13483 calories
Oatmeal	.05	15 cents... 9189 calories
Rice	.08	34 cents... 5086 calories
Beans	.05	14 cents... 7630 calories
Potatoes, at 75 cents per bushel	.01¼	22 cents... 7689 calories

W. O. ATWATER.

Century Magazine.
Reprinted by consent.

those which rank as delicacies are the costliest. If he uses the protein of oysters to make blood, muscle, and brain, it will cost him from two to three dollars a pound. In salmon, if he is enough of a gormand to buy it at the beginning of the season at $1 a pound, he will pay at the rate of $5 a pound for his protein. In beef, mutton and pork the cost of the protein ranges from a little over a dollar to about 40 cents a pound. (Salt pork, in which its cost is estimated at 25 cents, contains extremely little protein.) In such fish as shad, bluefish, and halibut (which are not mentioned in the diagrams), when they are cheap, say from 8 to 12 cents a pound, the protein costs about the same as in beef and mutton; but when the price is from 15 to 25 cents, the cost of the protein is from one to two dollars a pound. In cod and mackeral, fresh and salted, the protein varies from 30 to 80 cents a pound. Salt cod and salt mackerel are generally, fresh cod and fresh mackerel often, and even the choice fish, as bluefish and shad, when abundant, cheaper sources of protein than any but the cheapest kinds of meat. Among meats, pork is the cheapest; but salt pork or bacon has but very little protein and consists mostly of fat, which, though rich in potential energy, and very useful for people who have had hard work to do or are exposed to severe cold, is not so appropriate in warm weather or for those whose time is spent within doors and whose muscular labor is light. The comparative cheapness of cheese is well worth noting, and the great economy of oleomargarine as compared with butter deserves of more than a passing remark.

The comparison between wheat flour and potatoes is especially interesting. The protein in the wheat flour, at $6 a barrel or 3 cents a pound comes to 11 cents, while in potatoes at 50 cents a bushel it costs 15 cents a pound. Estimated in terms of potential energy, 25 cents pays for about 14,000 Calories in wheat flour at $6 a barrel, and 12,000 in potatoes at 50 cents a bushel. The potatoes would have to be reduced to 40 cents a bushel to make their nutrients as cheap as those of wheat flour at $6 per barrel. Adding to this the fact that the protein of wheat is the more valuable, weight for weight, because that in the potatoes is apparently less digestible and certainly of inferior chemical constitution, the showing against potatoes, even at this price, is very decided. But in the eastern portion of the United States, at any rate, people are very apt to pay 75 cents or $1 a bushel for their potatoes, while the finest wheat flour now sells at $6 a barrel; and if they are contented with flour of the coarser grades, they can have it for less.*

* At first thought this cheapness of wheat flour as compared with potatoes is a little surprising. The natural law of supply and demand of such staple materials, in the long run, shapes the price more or less closely to the actual value for use, and we should expect that the price of potatoes and flour would naturally gravitate to points which would make them more nearly equal in actual cheapness. At $10 a barrel, the price of wheat flour a few years ago, its protein would cost not far from 13 cents a pound, which would correspond to potatoes at about 60 cents a bushel. If the price of flour should remain where it now is, we may perhaps, expect that of potatoes to

In the United States the tendency to extravagance, combined with the mistaken notion as to the nutritive value of costly food, causes exception to the rule. Taking the world through, however, the poorer communities and classes of people almost universally select those foods which chemical analysis shows to supply the actual nutrients at the lowest cost. But, unfortunately, the proper proportions of the nutrients in their dietaries are often very defective. Thus in portions of India and China rice, in northern Italy maize-meal, in certain districts of Germany and in some regions and seasons in Ireland potatoes, and among the poor whites of the southern United States maize-meal and bacon, make a large part of the sustenance of the people. These foods supply the nutrients in the cheapest forms, but they are all deficient in protein. The people who live upon them are ill-nourished, and suffer physically, intellectually, and morally thereby.

Finally Prof. Atwater remarks that the Scotchman, as shrewd in his diet as in his dealings, finds a most economical supply of protein in oatmeal, haddock, and herring; and the thrifty inhabitants of New England supplement the fat of their pork with the protein of beans and the carbohydrates of potatoes, and supplement maize and wheat flour with the protein of codfish and mackerel; and while subsisting largely upon such frugal but rational diets, are well nourished, physically strong, and distinguished for their intellectual and moral force.

Mrs. Richards gives another way of stating the relative cost of some foods which is as follows:

Children and delicate women require food per day sufficient to furnish heat or equivalent energy equal to 2,500 to 3,000 Calories.

Sedentary men and more vigorous women require food per day sufficient to furnish heat or equivalent energy equal to 3,000 to 3,500 Calories.

---

come down gradually to a point where the actual expensiveness of the two will be more nearly the same. Of course this is a matter outside of chemistry, but the little study I have given it leaves me with the decided impression that the prices of such staple food-materials tend to adjust themselves to the nutritive values.

This statement is apparently in direct contradiction with a fact which these computations bring out most forcibly, to wit, the wide difference between the prices of foods and their values. But these differences have, really, a very simple explanation. The prices we pay for many of our food materials are regulated rather by their agreeableness to our palates than their values for nourishing our bodies. The sirloin of beef which we buy for 25 or 30 cents a pound is really no more nourishing than the shoulder which we get for 10 cents, or the neck at 8 cents a pound. In general, only a part, and often a small part, of what we spend for meats and sweetmeats goes for the nutriment they contain. The rest is the price of flavor, tenderness, and other things that make them toothsome. Nor does the disparity between animal and vegetable foods conflict with the principle I have ventured to lay down. Meats, fish, and the like gratify the palate in ways which most vegetable foods do not, and, what is perhaps of still greater weight in regulating the actual usage of communities by whose demand the prices are regulated, they satisfy a real need by supplying protein and fats, which vegetable foods lack. People who can afford it, the world over, will have animal foods and will compete with one another in the prices they give for them. These facts put the choicer animal foods outside the action of the law, if it be a law, that price and nutritive value tend to run parallel.

DIAGRAM IV.—POTENTIAL ENERGY OF FOOD.

CALORIES IN THE NUTRIENTS IN ONE POUND OF EACH FOOD-MATERIAL.

Beef, round, rather lean	807
Beef, neck	1108
Beef, sirloin, rather fat	1173
Beef, flank, very fat	2750
Beef, side, well fattened	1163
Mutton, leg	1142
Mutton, shoulder	1281
Mutton, loin (chops)	1755
Mutton, side, well fattened	1306
Smoked ham	1360
Pork, very fat	3472
Flounder	286
Cod	310
Haddock	331
Bluefish	474
Mackerel, rather lean	430
Mackerel, very fat	1026
Mackerel, average	696
Shad	750
Salmon	967
Salt cod	416
Salt mackerel	1364
Smoked herring	1343
Canned salmon	1036
Oysters	229
Hens' eggs	760
Cows' milk	328
Cows' milk, skimmed	176
Cheese, whole milk	2044
Cheese, skimmed milk	1166
Butter	3691
Oleomargarine	3679
Wheat flour	1655
Wheat bread	1278
Rye flour	1614
Beans	1519
Pease	1476
Oatmeal	1830
Corn (maize) meal	1616
Rice	1627
Sugar	1798
Potatoes	427
Sweet Potatoes	416
Turnips	139

The potential energy represents simply the fuel value of the food, and hence is only an incomplete measure of its whole nutritive value. Besides serving as fuel, our food has other uses, one of which is, if possible, still more important, namely, that of forming and repairing the tissues of the body, the parts of the machine.

W. O. ATWATER.

Century Magazine.
Reprinted by consent.

Hard-working robust men require food per day sufficient to furnish heat or equivalent energy equal to 3,500 to 4,000 Calories.

The average family, per person, probably requires food per day sufficient to furnish heat or equivalent energy equal to 3,000 Calories.

But foods differ not only in their power of furnishing heat or equivalent energy, but also in their cost, and since in general half the average earnings of the working man are spent for food it is extremely desirable to know what foods furnish the most heat or equivalent energy for the least money. The following table shows the cost of an amount of different food material sufficient to furnish heat or equivalent energy enough for the average member of a family per day. The amount of food required is assumed to be sufficient to furnish 3,000 Calories of heat or equivalent energy per day. Of course if the prices change the cost per day will change proportionately:

TABLE II.

SHOWING THE COST OF 3,000 CALORIES OBTAINED FROM DIFFERENT FOOD MATERIALS.

FOOD MATERIALS.	COST IN CENTS.
Suet at 6 cents a pound,	4.40
Potatoes at 30 cents a bushel, ½ cent a pound,	5.00
Corn Meal at 3 cents a pound,	5.43
Flour at 4 cents a pound, or $7.50 a barrel,	7.26
Flour at 5 cents a pound, or $1.50 a bag,	9.09
Potatoes at 50 cents a bushel or 1 cent a pound,	10.00
Sugar at 6 cents a pound,	10.41
Beef, from shin and flank, 4 cents a pound,	12.00
Sausage, Bacon and Ham, at 12 or 12½ cents a pound,	12.78
Beans and Pease at 8 to 10 cents a quart,	13.86
Sugar at 8 cents a pound,	13.92
Rice at 8 cents a pound,	15.69
Skimmed Milk at 2 cents a quart,	17.81
Parts of Beef, Mutton or Pork, pretty fat, 8 to 10 cents a pound.	20.00
Potatoes at $1.25 a bushel,	20.60
Skimmed Milk at 3 cents a quart,	25.62
Apples at 45 cents a peck,	27.30
Butter at 35 cents a pound,	30.74
Milk at 7 cents a quart,	34.74
Cheese at 14 cents a pound,	36.88
Green Vegetables at 5 cents a pound,	61.50
Beef, medium fat, with 15 per cent. bone, at 15¼ cents a pound.	100.00
Eggs at 18 cents a dozen,	106.50

THE SCIENCE OF NUTRITION. 57

From the above table it is clearly evident that suet, corn meal and flour, are at present prices, the cheapest kinds of food, but it must not be supposed that the above table teaches that we could live on single articles of food, suet alone for example, notwithstanding it would furnish the necessary energy and is cheap. With the exception of wheat, milk, eggs, and possibly one or two other articles, no single food contains all the elements in the right proportion. We need what are called nitrogenous foods, among which may be mentioned wheat, lean meats, pease and beans. We also need energy-producing foods, among which may be mentioned suet, butter, flour, corn meal, potatoes, sugar, etc. We also need mineral matter which is usually obtained in proper amount from the meats and vegetables. We also need water and air as well as flavors which make things taste good. The flavors are commonly developed during the cooking of the food, but we also often add flavors, such as vanilla, lemon, banana, etc. The air and water are from the common supply.

Lists or statements showing the proper combinations of different articles of food are called Bills of Fare, or more properly Dietaries, and much study is now being put on this subject.

For the most exhaustive study of nutrition in relation to cost and quantity and working power produced, see the pamphlet "Die Ernäbrung der Handweber im Zittan," von Dr. Carl von Rechenberg, Leipzig, 1890.

TABLE I.

AMOUNTS OF NUTRIENTS FURNISHED FOR TWENTY-FIVE CENTS IN FOOD-MATERIALS AT ORDINARY PRICES.

FOOD-MATERIALS.	At prices per pound.	Total food materials.	25 CENTS WILL PAY FOR:—			
			\multicolumn{4}{c}{NUTRIENTS, POUNDS.}
			Totals.	Protein.	Fats.	Carbohydrates.
	cts.	lbs.				
Salmon,	100	.25	.06	.04	.02	—
Oysters, 50 cents per quart,	25	1.00	.12	.06	.02	.04
Oysters, 35 cents per quart,	17.5	1.42	.17	.09	.02	.06
Salmon,	30	.83	.19	.12	.07	—
Bluefish,	10	2.50	.27	.25	.02	—
Beef, sirloin,	25	1.00	.29	.15	.14	—
Shad,	12	2.08	.29	.19	.10	—
Cod,	08	3.13	.34	.33	.01	—
Mutton, leg,	22	1.14	.34	.17	.17	—
Mackerel,	10	2.50	.35	.25	.10	—
Beef, sirloin,	20	1.25	.37	.19	.18	—
Mutton, leg,	20	1.25	.38	.19	.19	—
Beef, round,	18	1.39	.40	.29	.11	—
Canned salmon,	20	1.25	.44	.25	.19	—
Shad,	08	3.13	.44	.29	.15	—
Cod,	06	4.17	.45	.44	.01	—
Mutton, side,	20	1.25	.46	.17	.29	—
Beef, round,	15	1.67	.49	.35	.14	—
Salt cod,	07	3.57	.58	.57	.01	—
Salt mackerel,	12.5	2.00	.60	.30	.30	—
Mackerel,	05	5.00	.71	.51	.20	—
Butter,	30	.83	.73	—	.73	—
Milk, 8 cents per quart,	04	6.25	.74	.21	.23	.30
Salt cod,	05	5.00	.82	.80	.02	—
Milk, 7 cents per quart,	03.5	7.14	.84	.24	.26	.34
Cheese, whole milk,	18	1.39	.90	.38	.49	.03
Beef, neck, whole,	08	3.13	.92	.48	.44	—
Cheese, whole milk,	15	1.67	1.08	.45	.59	.04
Smoked herring,	06	4.17	1.21	.84	.37	—
Pork, salted, fat,	16	1.56	1.23	.04	1.19	—
Pork, salted, fat,	12	2.08	1.65	.06	1.59	—
Cheese, skim milk,	08	3.13	1.69	1.20	.21	.28
Wheat bread,	08	3 13	2.08	.28	.06	1.74
Wheat bread,	06	4.17	2.75	.37	.07	2.31
Potatoes, $1.00 per bushel,	01.7	3.24	3.04	.27	.03	2.74
Beans, 10 cents per quart,	05	5.00	3.96	1.16	.11	2.69
Potatoes, 75 cents per bushel,	01.25	18.00	4.13	.36	.04	3.73
Wheat bread,	04	6.25	4.15	.56	.12	3.47
Oatmeal,	05	5.00	4.48	.76	.36	3.36
Wheat flour,	04.5	5.55	4.83	.62	.06	4.15
Wheat flour,	04	6.25	5.44	.69	.04	4.71
Potatoes, 50 cents per bushel,	.85	26.47	6.06	.53	.05	5.48
Indian meal,	03	8.33	6.90	.70	.29	5.91

In the compilation of the twelve dietaries for thirty days which have been previously given with directions for their preparation, the following food values were the basis of the computation.

In this table the value is kept under rather than over the limit, and is intended to be available for any part of the country.

Article of Food.	Per cent. Proteid.	Per cent. Fat.	Per cent. Carbohydrate.		Calories per Pound.	Article of Food.	Per cent. Proteid.	Per cent. Fat.	Per cent. Carbohydrate.		Calories per Pound.
			Starch.	Sugar.					Starch.	Sugar.	
Bacon	10.00	70.00			2960	Ham	24.00	30.60			1650
Beans	24.40	1.50	51.	50	1650	Hominy	9 50	4.00	69.00		1650
Beef, shin and sh'ld'r.	20.00	4.00			520	Liver	20 00	5.00			560
Beets	1.26		0.10	8.00	160	Maccaroni	8.50	0.3	75.	00	1650
Butter	1.00	86.50			3615	Milk, whole	3.20	3.90		4.30	298
Bread	7.00	0.50	52 50	4.00	1150	Milk, skim	4.00	0.00		4.70	201
Cabbage	2.00	0.50	6.	50	190	Mutton, neck	15.50	8.50			619
Cheese	30.00	31.00			1780	Oatmeal	14.00	7.00	68.	50	1650
Corn Meal	9 50	4.00	69.	00	1650	Pork	2.60	77.80			3160
Codfish, salt	22.08	2.20			1488	Pease	2 00		21.	00	430
Corned Beef	21 40	18.00			1105	Pease	23.00	2.00	53.	00	1650
Corned Beef	2.60	77.80			2960	Rice	7.40	0.40	79.	00	1870
Crushed Wheat	11 64	1.26			1650	Sausage	13.00	40.00			1834
Fresh Fish	12.00	0.80			230	Squash	0 39		4.	00	160
Eggs	12.50	12.00	0.	55	705	Suet		89.00			3600
Flour, white	12.00	2.00	69.	00	1680	Sugar				96.50	1800
Flour, whole	11.65	1.30	71.	00	1650	Turnips	1.50	0.14	1.24	8.00	160
						Veal	19.00	3.00			462

Illustrative of this study of diet is given below the actual amount and variety of food eaten by a man and a woman on a certain day, with calculations of the amount of energy (expressed in Calories) contained in the food, together with the relative amounts of the different nutrients:

SUNDAY, FEBRUARY 22, 1890.

MAN.	BREAKFAST.	WOMAN.
Ounces.		Ounces.
4	Milk,	6
2½	Flour Griddle Cakes,	4
¼	Syrup,	½
⅛	Butter,	¼
¼	Cheese,	¼
3	Cream,	½
10	Coffee,	8
½	Sugar,	taste
11½	Oatmeal,	5
32⅛		24½

THE SCIENCE OF NUTRITION.

MAN.		DINNER.	WOMAN.	
Ounces			Ounces.	
3		Cold Corned Beef,	1½	
½		Fat Beef,	0	
5½		Vegetables, Parsnip, Beet, Turnip,	5	
11	1½ oz. waste	Baked Potato,	11½	2½ oz. waste
2		Bread,	0	
1		Butter,	1½	
Apple 3½	½ oz. waste	Fruit,	Orange 4	½ oz. waste.
1		Nuts,	¼	
0		Raisins,	¼	
59⅝			48½	

SUPPER.

5	1 oz. sugar	Pear sauce,	5	1 oz. sugar.
1¼		Bread,	1¾	
¾	⅛ oz. sugar	Cookie,	0	
¼		Butter,	¼	
1		Crackers at bedtime,	0	
0		Water,	12	
34		Milk,	6	
101⅞			73½	
5		Loss Waste,	5	
96⅞			68½	

Calculation of the above:

MAN.

Grms. Albumen.	Fats.	Sugar.	Starch.	Grms. Total.	
4.5	3.9	4.5		112.	Milk
2.5	10.1	2.5		84.	Cream
	37.			37.	Butter.
2.1	2.3			7.	Cheese.
7.3	4.6		45.5	91.	Bread.
28.6	6.3			84.	Beef.
		56.		56.	Sugar.
2.			53.0	252.	Potato.
2.3			15.	150.	Vegetables.
	14.			14	Fat in Beef.
10.5			35.0	70.	Griddle Cakes.
11.6	4.8		52.0	322.	Oatmeal.
				80.	Dry.
4.5	15.			28.	Nuts.
		17.		196.	Fruit { Apple. Pear.
75.9	98.		280.5	1,583.	Calories 2,343.

THE SCIENCE OF NUTRITION.

WOMAN.

Grms. Albumen.	Fats.	Sugar.	Starch	Grms. Total.	
13.4	11.8	13.4		336.	Milk.
0.4	1.7	0.4		14.	Cream.
	56.			56.	Butter.
2.1	2.3			7.	Cheese.
8.4	5.3		52.5	105.	Bread.
14.3	3.2			42.	Beef.
		35.		35.	Sugar.
2.0			53.0	252.	Potato.
2.1		14.		140.	Vegetables.
					Fat Beef.
16.8			56.0	112.	Griddle Cakes.
5.1	2.1		22.8	140.	Oatmeal.
				35.	Dry.
1.1	4.			7.	Nuts.
		17.		196.	Fruit { Pears. Oranges.
65.7	86.4	264.1		1,477.	Calories 2,130.

Since our American Standard dietary calls for 100 grams Albumen, 100 grams Fats, 450 grams Sugar and Starch, it will be seen that the amount was insufficient although the ratio of the weights 186 to 145 was about right: 1 pound of food for each 35 pounds of body weight.

The raison d'être of all this modern talk on cooking is that we may have better health; our civilization is being the death of us.

THE EVIDENCES OF GOOD HEALTH.

How shall we know if we are in our best condition?

First, we shall not be thinking about it at all. We shall not mind about the quality of our food very much. Life will hold other pleasures for us.

Mere motion, action, work, that is, use of muscular power, brings a delightful sense of life and force. The healthy workman goes to his day's work with vigor in his step, the schoolboy to his desk with eagerness.

If we find ourselves sluggish and tired in the morning it is because something is wrong. The standard of good health is for all alike the *consciousness of power*. We ask, How much *power* of *work* is there in the food we eat?—how much food do we need for a day's work? We call this *power*, *ENERGY*, and we reckon the force in Calories, that is in the mechanical equivalent of heat. This is the starting point of all our modern work in dietetics.

The modern standard, then, of good health is energy, power to do work; and by work we mean thinking, inventing, painting, writing, just as much as swinging a sledge hammer.

It is no longer a sin to be well and strong as it was in the days of the monasteries and hermits; man's ambition is no longer to be dyspeptic enough to see visions and dream dreams.

THE ESSENTIALS OF GOOD HEALTH.

What then is necessary to maintain this standard of physical strength? Good digestion waits on appetite. Exercise is the best sauce.

A sound firm has credit at the bank. A little pinch for money does not seriously disturb it; if one customer does not pay another does.

When a business house has to call on all it possesses day by day it is on the verge of bankruptcy. A sound man has a store of health, as it were, to fall back upon. He can bear cold and wet and hunger for a day or two, readily. When a little change in diet, a change in temperature or humidity seriously disturbs a man's health he is nearly or quite bankrupt.

E. H. R.

THE WHOLESALE PREPARATION OF FOOD.

So much has been done within a few years in the way of canning food that it would seem superfluous to consider any other method of preparation in the large way, but certain disadvantages inhere in these methods which render the question still an open one.

I. At present, canned food has been subjected, in order to kill all the agents of decomposition (the microscopic plants and their spores) to a temperature far too high or for too long a time, if the best flavor of the food is to be retained. Hence one tires of canned food.

II. The contact with metal injures the flavor of some things; glass may be too expensive.

III. The presence of solder in the can is a source of danger.

IV. The can once opened, the contents should be eaten at once, since such food is more liable to decay than that which is fresh.

V. The above requirements place most canned goods beyond the reach of persons of small means, and small packages have all the above mentioned evils intensified, beside the inevitable waste of material clinging to the dish.

In every city and town there is a mass of good food material practically going to waste because it costs too much time or too much knowledge to make it into wholesome or palatable dishes. To utilize it, some central station or kitchen should be established in which it can be prepared on well-proved principles and distributed by sale—daily, as bread, meat and milk are distributed.

In several foreign countries, notably Germany, "Peoples' Kitchens" (*Volksueche*) have been established. In 1890 there were fourteen *Volkskueche* in Berlin, in which were served 2,187,804 meals at noon, an average of 6,000 per day for 365 days, or 428 in each kitchen. Eighty per cent. of the meals cost 6¼ cents each and 4 per cent. cost 1¼ cents; the remaining 6 per cent. cost 3¾ cents each. The cost of running each kitchen was about $6,500 a year, but the oversight is done by volunteers.

The 6¼ cents portion contains the proper nutrition in the right proportions, and consists of a pint and a half of soup and three pieces of meat or fish weghing 6¼ ounces.

The dishes prepared are chiefly beans, pease and cabbage, with meats of different kinds.

The aim and results of these establishments are ably set forth in the account of the twenty-fifth anniversary of the starting of those in Berlin, just published. The following extracts are taken from Frau Morgenstern's paper:

"The greatest task before the civilized world of our time is the solution of the social question. The secret lies in the equalization of the different conditions of life by every possible elevation of the destitute and poorer classes, especially that of the day laborer. Advancing their material well-being improves the possibilities of their social and spiritual development.

" Before one can attempt to develop higher ideals for the advancement and happiness of the people, the food problem must be solved, then that of shelter. Only when the animal is satisfied can the spiritual man be reached. The hungry, the homeless man is unhappier among his fellow-beings than the wild animal of the woods who finds his home and food anywhere. Is it not a natural consequence when the hungry and the homeless become like wild animals, break through the laws of society and possess themselves through crime of what they want?

"Jacob Moleschott, the famous physiologist, in speaking of the moral effect of 'enough to eat' says 'Courage, good-will, and love of work depend in the highest degree on healthful, sufficient food—hunger lays waste the head and the heart.'

" For the working man only the best and most concentrated food is good enough."

But, as in many other things, while we may learn much from other lands, we should modify and adapt to suit our American conditions and to suit the spirit of our people and ways. Home and family life are our strongholds, the *cafe* living of Paris and the *Volkeskueche* of Berlin are alike foreign to our best ideas. We have even clung to the home manufacture of bread as no other nation on earth has done. We are slow to adopt any principle of coöperative living. The free American likes to be free in his selection of food, and preserves his individuality at the expense of his stomach as well as of his purse. This is the real reason why coöperative kitchens have hitherto failed; no two families like the same food or the same flavors. The problem, then, with us is a somewhat difficult one. The food must go to the families and not the people to the food, and only such dishes must be attempted as can have a somewhat cosmopolitan flavor, and such as can be easily prepared and will not suffer by being kept two or three hours, or will bear reheating.

The attempt to test the feasibility of such wholesale preparation of food was made in Boston, beginning in January, 1890, under the direction of Mrs. Mary Hinman Abel, and was made possible only by the generous financial support of Mrs. Quincy A. Shaw, to whom the scientific side of the question, as well as the far-

AVERAGE COMPOSITION OF SOME COMMON FOODS.

Nitrogenous.		Carbohydrates.			
Proteids.	Water.	Ash, etc.	Starch.	Sugar.	Fats.

Apples
Butter
Cabbage
Potatoes
Bacon
Milk
Bread (rye)
Bread (white)
Rice
Clams
Crackers (Boston)
Wheat
Eggs
Indian Meal
Beef, Average
Oatmeal
Mutton
Ham
Mackerel
Bluefish
Tripe
Fowl
Beans
Peas
Cheese

AN AVERAGE DAILY DIET SHOULD CONTAIN

Proteids...... .40 lbs.
Starch, etc.... 1.00 "
Fats.......... .40 "
Salts.......... .10 "
Total.... 1.90 lbs.

Computed and Drawn by Mrs. ELLEN H. RICHARDS.

reaching philanthropy, appealed most strongly. "The Story of the New England Kitchen," has been most ably and entertainingly told by Mrs. Abel in her report to Mrs. Shaw at the end of the first six months; now, at nearly the end of the second year under the able management of Miss S. E. Wentworth the success of the Kitchen is still more apparent. Branches have been established at 173 Salem street, Boston, and in Providence, R. I., in charge of Miss M. B. Gould, who will also open the one in New York city in October. It is expected that work at these branches will aid materially in solving some of the most perplexing social problems of the day.

The scientific aspects of the question of nutrition have been studied in connection with the daily work of the kitchen, for it is in such connection that the full value and significance of the results of scientific research can be best appreciated. For this research, a grant was received from the Elizabeth Thompson Science Fund, and also generous aid from Mr. Andrew Carnegie, his partner Mr. Phipps, and from Mr. Henry L. Pierce and others.

The results of this study are being slowly worked out. Much of the scientific thought embodied in this treatise, imperfect though it is, is the direct outcome of this work. Much material is still in hand.

It has been thought best to give the public the benefit of the knowledge, both in respect to success and failure so far obtained, rather than to wait for final conclusions, since it is only by combined effort under differing conditions that fixed principles can be established.

In no way can money be better made to serve the people than by securing to them good food and by teaching them to like it, so that they will be willing to learn to prepare it.

E. H. R.

THE ALADDIN OVEN.

WHAT IT IS.

It occurred to me one day that heat could be put into a box, kept there, and converted into work—the work of cooking.

What sort of a box?—an iron box?—no, iron will not hold the heat, it wastes it, and seems to cook the cook and not the victuals. The ovens of stoves and ranges, are iron boxes and are therefore not fit to be used.

Why should the iron boxes which make the ovens of iron stoves and ranges be ventilated? Because in order to cook food in them at all, such an excess of heat must be applied that they become fat boilers, or fat-rendering machines; the foul smells generated in this process are not wanted in any house.

What is the effect of this process on the food? This boiling or rendering is a process of partial distillation or dissociation; the fats are "cracked" as the chemists term it—the finer volatile parts and flavors are thrown off, generating unpleasant smells in the process, while the residue of the fat is left in a gross and indigestible condition with the other tissues which are deprived of their fine flavor.

That is the reason why it does not much matter what one calls for at many hotels and restaurants, all the meat tastes alike or is tasteless alike.

Why not, then, apply heat directly to the food in an oven like the common kerosene stove oven?

For many reasons: If the lamp smokes or smells the food is tainted; the direct heat scorches or burns without the slow penetration which is needed in fine cooking; the direct heat also dessicates the food and drives off the fine volatile flavors.

Conclusion.—All metal ovens,—all ventilated ovens—and all ovens in which excessive heat is applied to food, are more or less unsuitable to fine cooking. To that radical conclusion my observations have led me.

Having cleared away the obstructions by condemning almost every kind of apparatus now in use except, of course, a well-devised broiler to be used over charcoal—or an old-fashioned brick oven—or a tin kitchen before an open fire—the next question was:

What can be done about it?

Men have no right to scold their wives, or use swear words about the cook, and find fault with their meals in a constant and promiscuous way, if they only supply

THE SCIENCE OF NUTRITION. 67

them with apparatus to cook with that is not fit to be used; or which is so infernal in the heat that it generates, as to make it no wonder that those who have been of an angelic type and temper before beginning to keep house, should exhibit a capacity of another kind afterwards which may vex a patient man but ought not to cause him to complain.

The way out of this dilemma is for every boy to be put in the way of learning how to make first-rate bread, and to do all kinds of plain cooking, in one lesson of one hour by a little teaching in the simple principles. Girls may be taught at well, if they can spare the time from more important duties.

In the Aladdin Oven, the heat is put into an outer oven made of non-metallic and non-heat conducting material, which is, in fact, a form of stiff paper, made from wood pulp combined with other substances. Inside is a food receptacle nearly as large as the outer oven, made of sheet metal.

The heat passes around the thin iron wall of the inner oven through which it penetrates in even measure. This inner oven is closed so that the products of combustion and the direct drying heat of the lamp cannot enter it. It is provided with a ventilator which is used only in special cases.

THE ALADDIN OVEN.

WHAT IT DOES.

It cooks any and all kinds of food-material by processes corresponding to Roasting, Baking, Simmering, Stewing, Braising, Sautéing, Broiling, Grilling. It can be applied to making omelets or griddle cakes, and with a lamp or gas burner of high-heating power frying by immersion in very hot fat can be accomplished (but had better be omitted), both doors of the oven being kept open in order to give suitable attention to the process.

Breakfasts, for a family of eight or ten, can be prepared more quickly in all usual forms in this oven, by the use of a single lamp, than it can be when it is necessary to light a fire in the common stove or range.

The cooking of oatmeal, cracked wheat, hominy, soups, meat stews, and many kinds of fruit can all be done safely and thoroughly at night.

GENERAL INSTRUCTIONS FOR USE OF THE ALADDIN OVEN.

In adjusting the oven for use, care should be taken that the tube in the top through which a ventilating tube passes to the inner oven, is adjusted upon the nipple which is upon the top of the inner oven. This tube when properly adjusted rests flat upon the top of the outer oven and does not project, as in the first ovens made.

The lamp which is furnished with the Oven may be readily adjusted to heating the Oven by placing it underneath and resting it upon a block of wood or on a plate or saucer reversed, of such height as to carry the top of the chimney even with the under side of the outer oven. With this adjustment the maximum of heat will be conducted into the Oven and there will be no tendency to smoke from too close a contact of the top of the chimney with the metal. The lamp should not be put at the full height at which it may be expected to burn, for ten or fifteen minutes after it is lighted, lest it should smoke.

The oven will not become fully seasoned and will not do its work with full effect until about two weeks after it is first put into use. Usually the iron movable shelf which rests upon the bottom of the inner oven, keeping pans and dishes about half an inch away from absolute contact with the bottom of the oven, should be kept there in all processes of roasting, baking or simmering. When it is desired to broil, this shelf should be taken out. In a process corresponding to broiling, the large pan with the wire drainer in it should be put in. The slices of meat, chops or chicken should be laid upon this wire drainer, where they will be cooked in a manner closely corresponding to broiling.

For sautéing the pan should be put, with a little butter in it, directly upon the bottom of the oven. The cold Indian pudding, the fish, or whatever subject may be in process of sautéing, should then be laid in the pan and watched until it is done.

In order to toast bread, remove both the shelf and the pan; place the wire frame directly upon the bottom of the oven, lay the slices of toast upon the wire.

THE STANDARD ALADDIN OVEN has an inside space 18 inches in width, 12 inches in depth, and 14 inches in height; it is fitted with movable shelves so as to divide it horizontally into not over four compartments. An oven of *extra* size with a cooking space 21 by 13 by 15 inches is made on special orders. Cylindrical Ovens suitable for cooking for two to six persons are in progress, but the exact dimensions and prices cannot yet be determined.

LAMPS.

The lamp to be used with this oven may be either the Rochester, made by Edward Miller & Co., Meriden, Conn. ; the Gladstone, made by the Gladstone Lamp Company, 10 East 14th street, New York; the Daylight, made by the Craighead & Kintz Manufacturing Company; the Banner Lamp, made by the Plume & Atwood Manufacturing Company; the lamps made by the Bradley &

Hubbard Manufacturing Company; the Belgian American Lamp, made by the company of that name, 31 Barclay street, New York, the Pittsburgh lamp, or any other lamp of similar kind which has a circular wick about one and a half inches in diameter, with a central duct to convey air from below to the wick. In these lamps a practically perfect combustion is assured *provided they are kept in good order*. When carefully managed they may be worked either at full height or less, without smell or smoke.

If desired, the lamps will be furnished at cost to them by the makers of the oven, and will be packed in the oven; the oven is packed inside the table, from which it should be removed with care; the table should then be either on a stand, step, box, or table, from 12 to 18 inches in height.

When the *"full equipment"* is ordered the second table is furnished with the oven and the metallic table. If a lamp is used which is too high to be put under the oven, then the metallic table can be set up higher upon some blocks. If the lamp is too low for the top of the chimney to be about even with the bottom of the outer oven, then the lamp must set up on some blocks.

Inside the oven will be found one iron shelf with the edges turned down and without any holes in it; this shelf should be placed upon the bottom of the inner oven to keep the dishes a little off from actual contact with the bottom, so as to prevent burning at the point where the heat strikes. The other perforated shelves may be used or not, according to the number and height of the vessels in which food is to be cooked.

When many kinds of food are to be cooked at one time in the same oven, some of which may give off a good deal of water by evaporation, it will be expedient to have a dish or pan for the roast with feet to it about half an inch high, so as to keep the bottom of the pan from direct contact with the bottom of the oven. In this pan place the meat that is to be roasted, or any other kind of food to which it is desirable to give a brown or crusted appearance, then remove the close sheet of iron from the bottom and place these pans on the actual bottom of the oven itself; then put the close sheet of iron which has no holes in it on the middle bearings above the roast, and put the watery dishes on the upper spaces and if necessary open the ventlator a little so that the vapor can escape through the orifice.

By this arrangement a comparatively dry and browning heat will be attained in the lower space, while the moisture from the watery dishes will be wholly in the upper spaces.

Under these conditions, give about a third to a half as much more time as would be required in a common stove; with some kinds of food, twice as long. A little experience will be needed with each oven. Each oven will require a few days' seasoning in order to bring it to its normal condition and to overcome a little odor which is given off by the material with which the wood pulp is prepared.

If it is desirable to boil anything in a pot or jar, all the shelves may be removed and the vessel may be placed directly on the bottom of the oven, over the lamp; in this case the ventilator should be opened to let off the steam, but for ordinary work it is not necessary to open the ventilator.

The heat of this oven may be raised to about 300 to 400 degrees with the lamps described; it will be a little hotter at the bottom than at the top; but under this arrangement and at this degree of heat, fish, meat, custard, cauliflower and onion may all be cooked together without any flavor being imparted from one to the other; because there will be no distillation of the fats or juices of the food, only a little evaporation of water. At a much higher degree of heat there would be danger of the flavors passing over, and also danger that the smell of cooking might pervade the room, hence if a lamp of larger capacity and power is used it must be used with care.

A great deal of cooking may be done very slowly by night, by substituting a common flat-wicked lamp, of moderate power, for the kind previously named—wick one and a quarter inches to two inches wide—or a duplex burner. This lamp, if placed in position, will not raise the heat of the oven above 200° Fahr., at which degree of heat grain, meat and fish may be very slowly and tenderly cooked if left to take care of themselves throughout the night.

INSTRUCTIONS FOR HEATING THE ALADDIN OVEN WITH GAS.

There are many persons to whom the use of kerosene oil and the care of lamps is objectionable.

The oven can be worked with gas burned in a Bunsen burner as well as with the lamp, but it will require a little experience in order not to overheat, or to scorch or burn the food. One of my correspondents in Pittsburgh applies the natural gas with entire success and at a merely nominal cost. (See Appendix.) With the introduction of low priced gas for heating purposes, the economy of fuel applied to cooking may become as great as when oil is used.

If the common illuminating gas is used, apply the Bunsen burner, capable of consuming not over five feet an hour at the highest pressure to which the gas consumed on the premises can be subjected at night when the pressure is greatest. This will assure safety when the oven is worked by night.

The measure of gas required for ordinary work by day will be readily ascertained after a little practice.

BUNSEN GAS BURNER.

The best form of Bunsen burner is one which has a circular top with small perforations around the edges, each yielding a little tip of flame.

SPECIAL INSTRUCTIONS.

WARMING WATER.

Remove the tin tube in which the ventilator is, place a pan or kettle of water over the orifice and the heat from the oven will warm it sufficiently for many purposes; notably for washing cooking vessels, plates and dishes.

For this purpose it is not necessary or expedient to use hot water and soap. Put into a pan of tepid water a teaspoonful of kerosene oil. This oil has a great affinity for grease with which it combines in an emulsion. This process takes the grease from the pan or crockeryware without leaving any taint of either grease or oil.

Any one who is prejudiced against such use of kerosene oil may first try this process on greasy dishtowels: Put a little kerosene oil into a pan of tepid water; soak the towels in that emulsion, then pour it off and rinse out in clear cold water. No scouring required.

GENERAL DIRECTIONS.

Bread.—In baking bread which has been kneaded in the usual way, it is better to make the loaves so as to weigh from one and one-quarter to one and one-half pounds each; bake them two hours or longer according to taste, but change the position of the loaves when about half baked, placing those on the top shelf below, and those below on the top shelf. If it is desired to give the nutty flavor of crust to the whole loaf, the baking may be continued for a much longer period.

The Case bread-raiser is a most useful appliance; it consists of a wooden box with an opening in the bottom, into which slides a tin pan, which may contain water half an inch deep; the box has a glass front; about two inches above the pan is a perforated wooden shelf on which the pan of dough is placed to be raised; a little more than half way above is another similar shelf; a very small kerosene lamp with a flame about as big as a thumb-nail or a little larger, is placed under the bottom of the tin pan; this develops a moist heat by which the dough is raised ready for the baking pans in three and one-half to four hours, without regard to any outside conditions; it may then be removed from the raising pan to the baking pans, put back again into the bread-raiser for thirty minutes, and it is then ready for the oven. The advantage in the use of this bread-raiser is that the time can be established with certainty at which the loaves will be ready for the oven. The bread being made in the morning can be baked in the afternoon, or after the dinner has been removed from the Aladdin Oven. Orders for this bread-raiser may be sent to Mr. Daniel Dudley, 91 Carver street, Boston.

In order to make most excellent bread with the least work, knead with a spoon in proportion of one quart white flour, or whole wheat flour, or rye meal, with one pint or a little less warm water, salt to taste, and one-third of a cake of Fleischmann's compressed yeast (fresh) dissolved in half a teacup of warm water; stir ten minutes or more so as to mix the yeast thoroughly into a very thick batter like common dough, put in the pans and raise in bread-raiser about three hours, then move to the oven already heated and bake two, three or four hours, according to taste. (See subsequent reference to a bread-kneader.)

Roasting.—Place the joint or poultry in a pan, baste with butter and bread crumbs when put in, and place for the quickest work upon the tight shelf which rests upon the bottom of the oven, but not directly upon the bottom; for slower work, upon the next shelf.

Baking or Cooking Fish.—Baste with a little butter and crumbs of bread or cracker; place in a pan or crockery dish, and cook slowly upon the top shelf, or else cook slowly in a white sauce.

Imitation Broiling.—Have a broiling pan made containing a grill or perforated plate about half an inch from the bottom—such as is furnished with the oven when the order is given for a full equipment; put the meat upon this, and if you desire to work in the quickest way put the pan directly upon the bottom of the oven; if more slowly, upon the tight shelf. Cut the steak *two inches thick* if you want to have the most satisfactory results. Cook sausages in this way.

Braising.—Place the materials in a covered vessel; put on the cover, and put the dish as near the top of the oven as possible.

Simmering.—Use a flat-wick lamp of low power, and take all the time required.

Vegetables.—Nearly all vegetables, especially roots, require a higher degree of heat than meat or grain. Potatoes may be baked on the bottom of the oven, with the shelf interposed to keep them from scorching. Squash, cauliflower, onion, and cabbage, may be cooked in a satisfactory way, but it takes time, which each one must determine by experience. Asparagus, pease and beans may be cooked in the dishes in which they are served; in fact, all the work of the oven can be done in crockery or stoneware dishes, but for meats and large poultry it may be better to use ordinary baking pans. A separate table and separate lamp are recommended for boiling potatoes and for boiling water for service, when the cooking stove is not heated for warming the kitchen.

Soup Stock.—Put the materials in a cheese-pot and simmer all day or all night over a flat-wick lamp with wick one inch or one and a quarter to two inches wide.

Pastry and Cake.—If the tight bottom shelf will not hold all that is desired to bake, put a part there and a part above, and change about midway in the baking. Apple pies require two hours. Lemon pies, which are especially good, bake one hour and a half.

Game.—Venison may be treated like any other meat. Ducks, grouse, and the larger birds, may be basted with butter and bread crumbs, and roasted slowly and uniformly according to taste. Partridge and quail need no basting, but may be cooked in bread sauce, smothered in apples, or in any other way, at a moderate slow heat. Wild duck, grouse, partridge, and quail, placed in porcelain pots, with a white sauce and some French mushrooms added and cooked slowly, may give a new sensation even to a gourmand.

Tough meat and poultry may be rendered very tender, without being dried or scorched, by slow simmering for a long time; when tender, baste and roast for half an hour with the full lamp power.

Brown Bread, Pan Dowdy, and Beans, may be slowly cooked for a long time, and will be found to possess the flavors which are familiar only to those who can recall these New England dainties as they tasted when baked in the old-fashioned brick ovens, before anthracite coal and iron stoves had perverted all the old ways of preparing food.

I beg to add one more recipe for making bread which has been given me by a baker of very great experience:

"In answer to your questions as to the best yeast of domestic make for household use, I would say that the Vienna Compressed Yeast is most reliable.

"Suppose you try it thus: Take one ounce of yeast; dissolve in three quarts of water, warm as new milk ; mix in flour until as stiff as thick batter ; cover over top with sifted flour ; allow it to rise and fall flat once. Then add two quarts of water, cool as drawn from a well; dissolve in it five ounces of salt ; add flour to make suitable dough ; allow it to remain only long enough to become fairly light. Sponge requires about four hours; dough requires about two hours. Make up into loaves and put into pans, giving them about thirty minutes, more or less, as you see they require, before putting them into the oven.

"I think that if this method is followed and close attention given to it for a few times, it will be adopted hereafter. Of course, I assume that the yeast used shall be fresh."

INSTRUCTIONS OF A GENERAL KIND.

Dinner.—Place the sheet of iron that has no holes in it on the bottom of the oven so as to keep the dishes from absolutely touching the hot bottom itself; remove the next shelf, leaving the middle and upper shelves in their places. The dinner is to consist of four courses : First, soup ; second, fish ; third, roast; fourth, pudding. White or yellow ware dishes, common stoneware, or crockery vegetable dishes may be used. Two dishes or pans may be put on the lower shelf of the oven; two or three on the middle shelf; two or four on the upper shelf. Get dishes which fit well under these conditions.

The following dinner will call for a yellow or white ware dish, about two to two and one-half inches deep, for the meat; an oblong one with round corners is best. A tin pan can be used but it is not so easy to keep clean. The soup may be re-heated in a vegetable dish not too high to be set in upon the middle or upper shelf. Better heat it over another lamp on an iron table. A shallow dish will be required for the fish—one-half to two inches deep; two vegetable dishes for beans and pease, or for squash, tomatoes or onions, either two of which may be chosen. Lastly a pudding dish.

All the articles having been properly prepared, place either a sirloin of beef weighing six or seven pounds, a leg of mutton or lamb weighing six or seven pounds, or a pair of chickens of good size, in the meat pan. If chickens are chosen, stuff with soaked bread crumbs, seasoned with sweet marjoram, pepper and salt; baste the chickens with a little softened butter, and sprinkle with bread crumbs. If veal is selected, be sure to cook it long enough; give a little more time to white meat than to brown. Put three or four pounds of fish in the shallower dish, score it crossway, lay a few strips of pork on the scores, or else omit the pork and use a little soft butter and sprinkle with bread crumbs.

Put the soup made the day before, according to directions given elsewhere, into the soup dish, and season it to taste. A little catsup or celery seed in addition to the pepper and salt already in it, may suffice.

Place the potatoes ready to be baked in the tin pan. Put the string beans in one dish, pease in another, with a little water, just sufficient to cover them; add a little salt. If squash is selected, put it in a dish without any water. If tomatoes are selected, scald them and remove the outer skin; put them in a dish, add butter, salt and pepper, and sprinkle with cracker crumbs. If onions are chosen, put them in a dish with a little milk, salt and pepper. Be sure to give onions a long time.

Pudding.—Break up some stale bread into a soft mush with milk; add a little salt; grate in one-half a nutmeg, perhaps a little lemon juice, three or four great spoonfuls of sugar, one teacupful of seedless raisins or dried currants—raisins are the best; a little orange marmalade may be added. If you choose, omit the sugar and make a cold sauce of sugar, butter and a little spice mashed together. Do not be afraid of cooking the onions alongside the pudding. Old onions should be put in at the time the meat is put in. Young and tender onions require one and a quarter hours.

Suppose the dinner is to be served at 2.30 p. m. Put the meat or poultry in on the lower shelf at 12.30, so that it may cook two hours. Potatoes at same time if large; if very large, even earlier. All the rest may go in at 1.15 p. m.; then each dish will be taken out at about the right time.

Put the fish on the middle shelf, the tomatoes on the upper shelf, and cook them an hour to an hour and a quarter. The fish will be served first. Put the potatoes, if small, on the lower shelf, at 1.15 p. m., or much earlier if they are of large size; beans, pease, or squash on the middle or upper shelf at 1.15 p. m.; the pudding on the upper shelf at 1.15 p. m.; the soup anywhere to be reheated. Better heat it outside over another lamp. If heated in the oven it may be put in when the other articles are put in, or later. A little time will be taken up in placing the dishes, and they will come out in about the right order, the pudding being left in the oven while the rest of the dinner is being served.

It will be observed that the meat and potatoes take longer than the rest of the dinner, therefore they may be prepared first and put in as above. After the meat is in the oven, the rest of the dishes can be prepared and put in according to the above rules. When everything is in the oven, shut it up, see that your lamp is burning brightly and not smoking; set the dinner table, take off your apron and get ready to enjoy your dinner. If you do not want to be bothered, place the oven in the dining-room, serve from the oven to the table, and change the plates as may be required, placing the dishes and plates on a side-table behind a screen until they can be removed to the kitchen to be washed. Keep a very little fire in the kitchen stove to warm water if you do not use a kerosene stove of the common type for that purpose. In washing dishes use warm (not hot) water and a teaspoonful of kerosene oil to one pan of water.

If you want to make fine sauces learn how to make them from the books, in a blazer or chafing-dish, on the table. If you do not care to light the stove in summer, buy a common kerosene stove-lamp, and an iron table with a hole in it, upon which you may heat water, boil potatoes, and do other work of that kind.

This is the first lesson in plain cooking, and is in fact the only one required. In subsequent practice read some of the recipes in cookery books and reject nearly all the very complex ones; then apply common sense to those that you choose to try. It is not consistent with good cooking to disguise the fine natural flavors of meat, fish and fruit, with strong spices or other condiments; very strong flavors may be useful in order to disguise the poor quality of the food itself.

One of the great merits of this process of cooking slowly by moderate heat which does not distill the juice or dissociate the fats is, that the food when reheated or served the second time has no unpleasant flavor or greasy tang to it. In fact some kinds of food, such as veal or other white meats, seem to develop more flavor in the second process than in the first. It is a useful practice to cook such material for a second time in a blazer or chafing-dish, at the same time adding moderate quantities of Nepaul pepper, cayenne pepper, black pepper, tobasco or some other kind of sauce, curry powder, caramel, onion juice, bay leaves, sherry wine, etc., in order to comprehend the art of giving variety to the customary fare.

Very Slow Cooking.—Place oatmeal or cornmeal in porcelain jars, with sufficient salt, and somewhat less milk or water than would be commonly used.

Place meat scraps, bones, carcasses of chicken or turkey, corned beef, or smoked ham in earthen pots of sufficient size, with a very moderate amount of water according to what is wanted—whether stewed meat or soup. Salt and season according to taste; place in the oven at nine or ten o'clock p. m. Make use of a common lamp with a flat wick one and a quarter to two inches, or a low gas flame. In the morning the food will be found thoroughly cooked and the evaporation will have been very small. I have prepared twenty-four pounds of fresh meat, fish, oatmeal and cornmeal, with water and milk, in this way, and have found twenty-three and one-fourth pounds in the vessels the next morning. Ham, cooked in this way, should be afterward baked with a basting of bread crumbs, or may be cooked wholly by baking.

The method adopted in the New England Kitchen for making beef broth, for the sick or for the well, is as follows: (Samples may be found at 142 Pleasant street, where many useful ideas may be gathered.) A large tin vessel has been prepared which fills the inside of the oven. Into this twenty pounds of coarse beef bones broken up and twenty pounds of neck or shin cut in moderate-sized pieces are placed; to this material is added fifteen quarts of water, sufficient to cover well. This is placed in the oven at four o'clock p. m., and the lamp is then lighted. At seven o'clock p. m., the lamp is refilled, lighted, and left to burn itself out during the night, or in about eight hours. At six a. m., the broth will be found still nearly at the boiling point, about 204° F. The soup is then strained and cooled, and is ready to be put into jars for sale or seasoned for use. This broth is a nutritious and easily-digested food, differing from the ordinary beef tea which is mainly a stimulant rather than a food. The meat in very tender condition still contains much nourishment; it is chopped and seasoned so as to be eaten after or with the soup.

Each oven should be heated for about an hour before it is used, and each one may require a little experience in order to determine the time to which the various kinds of food must be subjected in it.

I have been asked how this apparatus might be applied in the best way to cooking or baking on a large scale. In my own judgment a Standard oven of the size described in this pamphlet, which can be worked with one lamp of the Rochester type and of the size in common use, may be adopted as the unit.

I can see no reason why these ovens might not be made in series, say of four or five, in one combination; each interior oven cut off from the next by a non-conducting wall, and each served with a separate lamp or gas burner. Each oven might then be heated according to the kind of food to be put into it; and there is no common cut of meat, and very rarely any turkey or other kind of poultry which this oven will not hold, together with some other dishes on the upper shelf. The extra oven may be worked with two lamps, and will hold very large poultry or joints.

I have a very large oven worked with three lamps, but I think it is cumbrous and inconvenient. Rating each Standard oven at about twenty pounds of fish, meat and vegetables at one time for one operation, a set of five would convert one hundred pounds in two or three hours; or if used for bread, ten to twelve pounds of bread to each oven; fifty to sixty pounds to the set of five ovens.

Tin or iron or other metal ware may be used to cook in, or, if preferred, stoneware and common chinaware may be safely used to cook in, as the heat of the oven is not sufficient to harm them.

Food when cooked in stone, china or agate ware, may be served upon the table in the same dishes.

Persons who feel any uneasiness about using kerosene oil, which is the cheapest source of heat, may obtain gas burners made on the Bunsen principle, which can be applied with a rubber connecting tube, at E. H. Tarbell's, No. 111 Washington street. The use of the burner calls for some practice lest the oven should be overheated and the food scorched or burned. I purchase my kerosene oil of the 150° F. standard, from Allen, Bradley & Co., Central Wharf, Boston, about 30 gallons at a time.

Since the foregoing was written the oven of extra size has been perfected, in which large joints or poultry and other dishes for a dinner for a large family may be dealt with by the use of two lamps. Whether or not it will be expedient to go beyond this extra size remains to be determined by future experiments.

Cylindrical ovens of two sizes—one suitable for three or four persons, and one for five or six—have been substantially perfected. Litigation—which has prevented the use of the burner most suitable to the proposed workman's pail—has been settled, and I may soon hope to complete that appliance, which has been the objective point of all my work.

THE ALADDIN OVEN.

HOW IT DOES IT.

Since my first general instructions were given, I have been asked many times "When my cookery book would be published?"

Even now I can only give the results of observation and of partial study with the recipes which have been prepared at my own suggestion, coupled with others that have been sent me. For more systematic rules and recipes see Part II.

Since the first edition of these instructions was issued, much experience has been gained and the work which can be done in the Aladdin Oven has been very much extended. I have made arrangements to heat the kitchen in my winter house mainly from the furnace. Since it is no longer necessary to get up a fire in the early morning, my cook finds that the breakfast can be prepared more quickly and with less work in the Aladdin Oven than upon the range. Even if the range or stove must be lighted for other purposes, the Aladdin Oven is ready for the work before the oven in the range is in suitable condition for cooking. Moreover, since it is no longer necessary to force the range, a great waste of fuel is saved, especially if a little coke is used either by itself or in starting the hard coal fire.

I can hardly doubt, however, on the basis of my own experience and that of many others, that the use of the iron stove or range will almost wholly give place to the oven. There is no question that the fine flavors of meat, fruit, grain and vegetables are developed by the application of heat at the right temperature. These flavors are very subtle and very volatile and are doubtless due to chemical changes. A high heat dissipates these fine flavors about as fast as they are formed, especially when the cooking is done in open vessels, while a moderate heat which only evaporates some of the water in the material develops the flavor without sending it off in vapor, especially when the work is done in closed vessels. A high heat converts fat into a fatty acid which is noxious—a low heat prepares the fat for ready assimilation in the process of nutrition in due proportion.

True food is not prepared by merely making a cook's mixture commonly called a recipe—it is a fine process of the chemical conversion of crude food material under the controlled and regulated application of heat.

In preparation for breakfast, prepare

COFFEE, SOUTHERN METHOD.—Grind the roasted coffee very fine; put it into a pitcher at night, pour cold water upon it equal to the quantity of coffee expected to be made; let it steep all night. In the morning strain it, and, if desired, clear it with the white of an egg; then heat to the desirable point for serving. It can be heated in the oven, or over another lamp, or on the stove.

Having lighted the lamp at once on coming down in the morning, so as to heat the bottom of the oven while the food is being prepared, a process corresponding to the work of frying in a spider by the use of a little fat or butter, without the complete immersion of the food in the hot fat, may be applied to a great many kinds of food. The method is as follows, scientifically called

SAUTÉING.—Place a tin or metal dish immediately upon the actual bottom of the oven, directly over the lamp. Put a little butter or lard into it. Having some cold mush made of cornmeal or of oatmeal already very thoroughly cooked, cut in thin slices, put them into the pan, leave until brown on one side, then turn and brown on the other side; serve very hot. Pan fish may be cooked in the same way. Try almost anything in this way. Crisp soda biscuit may be salted and cooked in butter in this way; they make a delicious side dish.

GEMS.—Mix Graham, rye, or whole wheat flour with water and salt into a thick batter, using neither yeast nor baking powder. Having placed a metal biscuit pan in the oven so that it is already well heated, put in a little butter, merely enough to keep the gems from adhering to the metal; bake quickly on the bottom of the oven.

The term "whole" or "entire wheat" has been misapplied to the flour which is called by that name. This flour is made from grains of wheat from which the outer cuticle has been removed. Graham flour is made from the whole grain, including the outer covering. Preference is given by many persons to the Graham flour. Care should be taken to secure a special brand of Graham flour made from sound wheat, as the common grades bear the ill repute of being a compound of the rubbish of the flour mill with the poor wheat. These common grades may, therefore, contain more bran than anything else.

(See Appendix—Letter from Mr. Louis H. Hyde.)

I will first give a few of our own recipes, mainly for dishes suitable for breakfast.

BROILED CHICKENS OR BIRDS.

Place in a tin pan slices of cold toasted bread, place upon them a little butter, put over the toast a drainer, that is, a wire frame or a tin plate pierced like a colander, half to three quarters of an inch high. Upon this drainer place the chicken quartered or the birds split for broiling. Very thin slices of pork may be laid over them, or else smear with a little soft butter and sprinkle with bread crumbs.

BEEF STEAK AND MUTTON CHOPS.

Place the drainer, the one made of wire being the best, in a tin pan. Put a little butter into the pan, place the meat upon the frame, thick slices preferred; cook according to the thickness of the meat.

SAUSAGES.

Proceed in the same way, omitting the butter. Put the lard oil which is drained out of the sausages into the grease pot, *then* eat the sausages.

FISH.

Proceed in the same manner, adding pepper and salt according to taste. Put a little butter upon the fish and sprinkle with bread crumbs, or cut the fish cross-way and lay in thin strips of salt pork, then sprinkle with bread crumbs.

HAM OR BACON AND EGGS.

Cover the wire or tin drainer all over with thin slices of ham or bacon, place in the oven and cook fifteen to twenty minutes, the oven having been already heated. Then break the eggs upon the meat, return to the oven and cook five to ten minutes according to taste.

When very quick work is necessary, all the dishes above named may be put directly upon the bottom of the oven over the lamp. They must however be watched, as it is possible to scorch food which is cooked under these conditions without the intervening shelf that keeps the bottom of the pans from the direct contact with the bottom of the oven.

When very quick work is not required, leave the close iron shelf which has no holes in it upon the bottom of the oven, and place all the dishes upon that, except the imitation fry. Sautéing and grilling must be done on the actual bottom of the oven. Slow work is the safest. The longer meats are subjected to a moderate heat without burning or drying up the juices, the more tender and digestible they will become.

BISCUITS.

Biscuits made of any kind of meal may be baked in biscuit pans upon the shelf next above the bottom of the oven. They may be baked while meats and fish are being cooked below. Heat the biscuit pans first.

GRIDDLE CAKES.

Take out the close iron shelf, place the griddle in the oven on the bottom as soon as the lamp is lighted, so as to heat it with the oven; put the lamp, well trimmed at full power and proceed in the usual manner to cook griddle cakes, pan-cakes or flap-jacks.

MUSH OR PORRIDGE.

Made of cornmeal, oatmeal, cracked wheat or hominy; place either in porcelain vessels; add salt, moisten sufficiently, but not quite as much as if the work were to be done on the top of the stove where the water would greatly evaporate; cover the pot and put it into the oven for many hours; all night if you please.

This slow cooking at moderate heat will not injure but will develop the fine flavor of the grain. This mush or porridge may be eaten in the usual way or may be allowed to become cold; then cut in thin slices and cook in the pan in imitation of frying, according to the instructions previously given.

ROASTED WHEAT CAKES.

If grains of wheat are roasted as coffee berries are roasted, they can be ground in a French coffee mill which should not be used for any other purpose. This brown flour can then be made into biscuits or into griddle cakes that possess a peculiar flavor which many people like. It must be very digestible. Coffee berries or wheat grains may be roasted in the Aladdin Oven. It takes a long time, but requires very little attention.

It therefore appears that almost all the customary breakfast dishes, except those which require frying by immersion in hot fat, can be prepared in the Aladdin Oven. The ovens stop short at the doughnut stage.

PAN-CAKES.

One pint of milk, four eggs, a little salt, two heaping tablespoonfuls of flour. Beat the eggs and stir the milk and flour together. Heat the pan on the bottom of the oven ten minutes; then butter the pan, pour the pan-cakes in, cook for twenty minutes, turning once midway.

OMELET.

Heat a tin pan upon the actual bottom of the oven, put a small piece of butter therein; beat up the eggs and stir into the pan, place it upon the bottom of the oven for about four minutes, closing the door; then turn over and serve at once.

The next directions apply to the preparation of the more substantial dishes.

ROASTING.

Give more time than in the common oven and do not be tempted to try to roast quickly.

BROWNING MEAT AND POULTRY.

Some complaints have been made and exceptions have been taken for the lack of an æsthetic appearance in the roasts. Bread crumbs and butter are turned brown by heat at a much lower degree than the fats of the meat itself. Therefore smear the joint or sirloin with soft sweet butter; sprinkle with bread crumbs when the dish is put into the oven. This will give an æsthetic appearance that will be wholly satisfactory.

BROWNING HALIBUT AND CUSK A LA CREME.

Having prepared the fish in the usual way, well seasoned, with sauce, wholly omitting the customary boiling by which the fine flavor of the fish is commonly destroyed, cook the fish a sufficient time according to the quantity; about half an hour before serving beat the white of an egg into a froth and spread it over the fish. The result will please the critical eye and it will not hurt the fish.

TOASTING BREAD.

Take out the lower shelf, place a wire grid directly upon the bottom of the oven; then lay the slices of bread thereon.

BAKED BEANS, BROWN BREAD, INDIAN PUDDING.

Many of the younger generation never really tasted either of these pre-historic dishes. The following recipes are therefore given:

BAKED BEANS.

Soak one quart of beans in water all night; turn off that water if you don't want too much of the bean flavor and add fresh water. If you are a true bean eater, do not turn off the water, but put the pot in a heated oven and leave it there one hour; then draw off the loose water. Place half a pound of salt pork on the top; add a teaspoonful of salt; put back into the oven and cook the beans all day and all night if you want to. Add a little molasses if you want a little sweetening. Beans can hardly be overcooked.

BROWN BREAD.

Two cups of Indian, that is, maize meal; two cups of rye meal; one and one-half teaspoonful of baking powder. One-half teaspoonful of salt; one small cup of molasses (treacle). A little over a pint of skim-milk. When baked in the Aladdin Oven, the dish should be put on the shelf next above the bottom of the oven, where it should bake three hours with the lamp at its full capacity.

INDIAN PUDDING.

Two quarts of milk; one-half cup of molasses, one cup of Indian meal. One-half teaspoonful of salt; piece of butter. Bake in the Aladdin Oven four hours on the top shelf with the lamp low, in rather a flat dish.

NOTE.—Cornmeal, "country ground," so-called, that is, cornmeal which has been ground by slow processes, and which has not been subjected to drying in kilns, is much better and possesses a much higher flavor than the common cornmeal, such as we find in the shops. I attribute this difference to the fact that the meal has not been heated; therefore its fine flavor has not been driven away. Meal of this description, both of the Rhode Island white variety and the common yellow meal can be purchased of Messrs. John B. Chace & Son, 42 South Main street, Providence, R. I.

FISH AND CLAMS, SEASIDE FASHION.

Take out the shelves from the oven; make use of a tin box of sufficient size and depth; in the bottom put some wet rock-weed or sea-weed; put in the clams; add another layer of rock-weed; put in the fish; cover with rock-weed; put the cover on the box, if you have a cover; put into the oven and subject to heat for a sufficient time. This method corresponds to a clam-bake.

SAM WELLER HASH.

Chop into pieces of a rather large size, one pound of ham and two pounds of veal; add cold water and seasoning; and, if you like it, thicken with a little bread-crumb or flour; simmer slowly a very long time. After being sufficiently simmered, this compound may be put into a baking-pan, covered with a light crust, and converted into a "weal and ham" pie.

FISH CHOWDER.

This dish can be made in perfection in the following way: Cut the pork in thin slices, place in the dish which is fitted with a drainer, put this upon the bottom of the oven, try out the fat, and then make use of pork in the usual way in making the chowder, combined with fish, milk, potatoes, crackers, onions, etc., according to the customary method. Simmer slowly with moderate heat.

It will be observed with respect to chowder and many other kinds of food which when cooked at a high heat are not appetizing if heated over again, that when cooked at a moderate degree of heat they may be warmed over the next day and will prove to be quite as appetizing, and in some cases more so than when first prepared.

SALT FISH.

Any one fond of this dish may prepare it in the best way by moderate cooking in a dish with a little water added, preparing the pork scraps in the same way that is directed for trying out the fat of the pork for making chowder. Cook the beets thoroughly.

COOKING WHOLE HAMS.

Soak the ham first. Use a lamp of moderate power ; put the ham in a cheese-pot, add half a bottle of cider or red wine; fill up the interstices with sweet meadow hay ; very little water to be used ; put on the cover and cook very slowly for a long time according to the size of the ham. Remove from the pot, take off the skin, baste with bread crumbs ; do not use cloves unless you like them ; place a high-power lamp under the oven and roast on the lower shelf half an hour.

CHEESE.

Reference may be made to the chapter in Dr. Mattieu Williams' Chemistry of Cookery on the nutritive value of cheese provided it is thoroughly cooked. An excellent pudding may be made in the following way : Two parts stale bread ; two parts skim milk ; one part skim-milk cheese broken up into fragments; season as for a Welsh rarebit ; add a very little bi-carbonate of soda, according to the instructions given in Dr. Williams' book. Cook for a long time at a moderate heat. To those who are fond of cheese this dish will prove to be appetizing and digestible. The proper cooking of cheese converts it from an indigestible substance into a very nutritious one.

Persons who have preferred very rare meat or game may find the flavors more fully developed if the process of cooking is continued so that the color is more brown than that of the rare part of the joint cooked in the ordinary method, while at the same time the outside will not be dried or scorched.

A NEW METHOD OF PREPARING "HOG AND HOMINY."

It is well known that the favorite dish of the Southern black man is "hog and hominy." The combination appears to be readily digested. It is said that bacon and wheat meal may not be easily digested, or beef and cornmeal. The "hog and hominy" is a very strong food. There is probably no body of workmen subsisted at so low a cost, regard being given to the force generated in the food, as the black men who work on a ration of bacon and corn meal. The customary allowance in the South is, I believe, three and one-half pounds of bacon and one peck of corn meal for each week's supply. A very delicious compound may be made by placing in a pot one cupful of meal previously moistened with water sufficient to swell it without softening it too much. Upon the top lay a few sausages ; put in the oven and cook at a low heat for five hours. Bacon may be cooked in the same way, but the meal requires more water. This is one of the dishes that I propose to put into the workman's pail, with a view to the lamp being lighted when the work begins ; the dinner to be ready five hours later at noon.

SALT CODFISH AND CREAM.

Break the fish into slivers ; put it into a porcelain or earthern pot with cream, if you have it—or with milk or a little butter added ; cover the vessel and cook at a moderate heat for one hour or more ; the longer the more tender.

PASTRY.

Pies may be baked in the most perfect manner in the Aladdin Oven. The use of tin rings with separate circular plates pierced with many small holes for the bottom, is recommended in place of the common tin plates. The bottom crust will then be as absolutely and as thoroughly baked as the top crust. "*Pale pie*," may thus be banished from the customary bill of fare.

By permission of Mr. James F. Case, I give a recipe for what is named

"CASE'S HEALTH BREAD."

For five loaves of one and one-Half pounds each, two pints of oatmeal gruel previously cooked very thoroughly ; three and one-half pounds of "whole or entire wheat flour"; one-half pint of milk ; one tablespoonful of sugar or two teaspoonfuls of New Orleans molasses ; one teaspoonful of salt ; one cake of fresh Fleischmann's Yeast. Use the oatmeal gruel and milk for wetting the flour. Knead thoroughly. Keep in the bread-raiser until the batch has about doubled itself in bulk ; then form into loaves ; put back into the bread-raiser for thirty to forty minutes. Then bake in the Aladdin Oven, previously heated, for about one hour and a half.

In my own household and office practice we have made use of even a larger portion of oatmeal, and have also mixed Graham flour with it in place of the whole wheat flour. A very delicious kind of bread may also be made in the Case Bread-Raiser and the Aladdin Oven almost wholly of rye meal ; only enough wheat flour being dusted upon the hands, if people will persist in kneading bread by hand, or upon the spoon, to overcome the stickiness of the rye. Even hand kneading with a spoon may be displaced by a very simple mechanical spoon or Bread-Kneader sold by Edwin Prescott, 21 Hamilton street, Boston.

Any one who never cooked anything before may begin in the following way : first, by making some bread, which is commonly considered a great mystery but is really a very simple matter. Let it be assumed that the oven has been brought into a seasoned condition by being heated a few hours each day for a week.

A beginner may well learn how to make bread as a first lesson, I therefore give the following definite instructions :

BREAD.

Mix in a bowl one quart of flour, either white, whole wheat or rye meal, with a little less than a pint of water at blood heat—90° F. Add one half a (level) teaspoonful of salt ; mix one-third of a cake of Fleischmann's compressed yeast in one-half teacupful of warm water; be sure that the yeast is fresh. Add this dissolved yeast to the flour and salt; stir with a spoon into homogenous dough. The whole object of kneading with the hand, the spoon or the mechanical kneader is to diffuse the yeast evenly in the dough. I have found it a very simple matter—easily mastered after one or two trials. Butter the bread pans and put in the dough to rise. If you have a Case Bread-Raiser, use it for the raising of the bread according to the instructions given with it, for about three hours.

Time the work so that the pans may be transferred to the oven when it is well heated, perhaps after the dinner has been removed from it. Then bake two hours; a little less or a good deal more, according to the taste. If you have no bread-raiser place a pan of warm water on the lower shelf of the oven; place the dough in the bread pans on the middle and upper shelves. Light the lamp, and keep the flame as low as possible, yet burning. Close the oven, and in about three hours the dough will be raised. The important point is to keep the dough in a humid atmosphere at 90° F. for a certain time—about three hours. Open the oven, quickly remove the pan of water, place the bread pan upon shelf which rests upon the bottom of the oven, turn the lamp up to its full height, and bake about three hours. It may be safer to knead the bread in the usual way. Ask the cook how to do it; then raise it and bake it according to the foregoing rules.

By the kindness of Sir Henry Thompson, M. D., I may add two recipes which he prepared for London bakers many years ago, for bread which he recommends in his little work on "Food and Feeding":

DIRECTIONS FOR MAKING WHOLE-MEAL BREAD. (IN FLAT CAKES.)

WITH BAKING POWDER.—Take two pounds of coarsely-ground whole wheat meal, and half a pound of fine flour, or, better still, the same weight of fine Scotch oatmeal. Mix thoroughly with a sufficient quantity of baking powder and a little salt; then rub in two ounces of butter and make into dough—using a wooden spoon—with cold skimmed milk or milk and water, soft in consistence, so that it can almost be poured into the tin ring, which gives it form when baked. In this manner it is to be quickly made into flat cakes (like tea cakes), and baked on a tin, the rings used being about an inch high and seven or eight inches in diameter, each inclosing a cake. Put them without delay into a quick oven at the outset, letting them be finished thoroughly, at a lower temperature.

WITH YEAST.—When good German or other yeast can be obtained, add the necessary quantity to the dough, made as above directed with the two meals, butter, salt, and warm milk and water; make the cakes and put them on the tin with their rings, and set near the fire to rise, which they will do in an hour or little under. Then bake in a medium oven in the same way as for any other fermented bread. When yeast is used and not baking powder, a medium coarse oatmeal may be added to the wheat meal.

The object of making this bread in flat cakes or in scones, is to insure a light and well-cooked product. It is difficult to insure these two qualities in the form of loaves except of the smallest size. A larger proportion of oatmeal, if preferred, can be adopted by either method.   H. T.

Sir Henry Thompson adds in his note to myself the following :

"One reason for using oatmeal also, is, that not only is a more nutritious combination made, but a less soft and spongy mass when baked, i. e., in the interior of the bread ; but the mixture is much better when cooked in flat cakes one and one-fourth inches to one and one-half inches thick ; about seven or eight inches in diameter. The flavor, too, is excellent. On my own table I use one-third oat and two-thirds wheat meal."

PORK CHOPS SMOTHERED IN APPLES.

Place sliced apples in a dish ; lay the chops upon them ; sprinkle with a little salt ; cover again with sliced apple ; put a cover upon the dish, and cook thoroughly, more or less time according to the quantity.

Chicken, quail and some other kinds of meat, mainly white meat, may be cooked in the same way with very appetizing results.

CHICKEN, QUAIL AND PARTRIDGE, FRENCH MUSHROOM OR CELERY SAUCE.

Cut up the chickens, halve the partridges, cook the quail whole. Place in dishes which may be covered ; make a white sauce of bread crumbs ; season it to taste, but mainly depending for flavor either upon French mushrooms and the liquor which comes with them, or upon the root or hock of celery previously simmered so long that it can be reduced to a purée; make a sufficient quantity of sauce, to fairly fill the dish when the birds are put in, and simmer slowly with the cover on, for such a length of time as may correspond to the quantity.

MINCED GANDER ON TOAST.

Order the toughest old gander that can be found in the market ; prepare him in the usual way ; stuff with onion stuffing, or if preferred, in European fashion with prunes and chestnuts. Place in a large vessel with a moderate amount of water, the vessel being covered ; place in the oven ; light the lamp at about half power, and simmer slowly all night ; test with a fork in the morning, and if not tender, add a little more water if more is needed, and, simmer all day. In about eighteen hours the work will be done. The meat will then be so tender that it cannot be carved ; mince and serve on toast with plenty of gravy, of which there will be an abundance.

FOWLS.

Select a pan about an inch and a half deep that will hold two or three fowls. Put the fowls in the pan, stuffing them if you like. Lay thin slices of pork over the fowls. Fill the pan up to the lip with water. Put in the middle of the oven and simmer four or five hours according to the age and size of the fowls.

The fowls may then be removed from the first pan ; may be basted with bread crumbs and butter, and served as a roast ; or may be served as boiled fowl, to which they are far superior, with a butter and egg sauce. The water in the pan should be saved, allowed to cool ; remove the grease, and use it the next day with the carcasses to make broth or soup.

MACARONI.

Macaroni may be cooked without being previously boiled, with any kind of sauce that may be preferred. The dish which is very popular with my own friends, consists of macaroni moistened with liquor from the can of French mushrooms, the mushrooms being also cooked with the macaroni; no cheese. For appearance add a little butter and bread crumbs, sprinkled over the top.

DUCK AND GROUSE.

When a distinguished purveyor was asked what to do with the carcass of the duck after the breast had been removed, he replied, "Better give it to the poor." This is not wholly true. The leg of the duck when tender is very good, but there may be a better use for the legs and wings either of duck or grouse, which are apt to be wasted when cooked on the bird. Remove them before cooking, place them in a pan with a little water and seasoning. Simmer them a long time until all the flesh can be taken from the bone, then the lean part may be forced through a colander, or else mince very fine and use with the giblets in a gravy or sauce. Roast the rest of the duck, more or less according to taste, in the usual manner The abundance of sauce made in this way will be appreciated.

A Washington method is to cook coarse hominy and place it with the duck so as to absorb the juices. Hominy should first be cooked separately; it would probably be better to simmer it very slowly and for a long time; then use it. Cooking hominy is claimed to be a fine art, but it is done in perfection in the Aladdin Oven.

WASHINGTON DUCK. (WITH ADDENDA.)

Place in a granite-ware or crockery dish, heating it first in the oven, a layer of very coarse hominy, rice or macaroni, with as little water as will serve to soften it in the process of cooking; place upon the bottom shelf and cook until it is about half done.

Make a thin batter by beating a little butter, bread crumb and egg together; split chicken, grouse or partridge as if for broiling (legs and wings may be left on or may be removed to be simmered separately), small birds whole—roll in the batter, add a little butter and salt to the hominy, rice or macaroni, place the birds thereon; return to the oven on one of the middle or upper shelves and leave until done.

For variation, add to the hominy, rice or macaroni, a few fresh mushrooms, or canned mushrooms and liquor—with chicken add some curry—with duck a suspicion of cayenne and onion—a little jelly sauce—or chop some celery and cook it from the start with the other base. With chicken a couple of sausages may be added.

The gravy and fat will be absorbed either on the hominy, rice or macaroni, and the dish will be most appetizing.

Another way is to put the hominy, rice or macaroni and the bird into a dish which has a close cover, cook thoroughly before removing the cover; then beat up a little bread crumb and butter with the white of an egg, spread this over the surface, remove the cover and put the dish upon the upper shelf where it will be well browned. If the birds are preferred rare, keep them out until half or three-quarters of an hour before serving. Use gumption.

If the legs and wings with the giblets are to be simmered, start cold so as to draw all the juices, in order to make gravy—the meat may be very finely-chopped and added—or if simmered long enough, converted into pulp or purée.

CRISPY CAKE.

Most of the labor which is expended in rolling thin gingerbread and other kinds of cake, is wasted. The common idea that the crisp quality of thin cake depends upon a very hot oven, is a blunder. The proper method for cooking thin and crisp kinds of cake is as follows:

Use either a large tin baking pan wrong side up; or, what is better, a sheet of Russia iron which it is better to have prepared with the edges turned over all around so that it may not spring in service. Spread what may be called the dough or mixture that is to be cooked, very thin, smoothing and spreading it with a knife without previously rolling it. Leave the tight shelf upon the bottom of the oven, and place the pan wrong side up on the sheet with the cake spread upon the bottom, on this lower shelf. Bake, as a rule, about forty five minutes. But every one will want some more. Enough of such cake is always a little more. The area of the bottom shelf of the oven is limited. I have had some large sheets of iron prepared; four to lay flat directly upon each shelf; four others with the edges turned down about an inch and a half, so that I can get eight sheets into the oven. These may be put in, and the bake may begin while they are all in the original positions, one above another, but they must be changed in succession so as to get each sheet upon the bottom for a little while, if it is thought expedient to give all a uniform brown color. If the whole number of sheets is made use of, the product will be about twelve square feet, which may possibly be enough, although a little more will soon be called for.

The recipes are as follows:

THIN SUGAR GINGERBREAD.

One cup of flour, one cup of sugar, half a cup of butter, one teaspoonful of ginger, one egg, three tablespoonfuls of milk. Beat the eggs and milk together; add the other ingredients. This is the unit for a small quantity.

THIN MOLASSES GINGERBREAD.

One cup of flour, half a cup of butter, half a cup of sugar, half a cup of molasses, one teaspoonful of ginger, one egg. Beat the eggs and milk together and proceed as in the previous recipe.

BROWN PUDDING.

One cup of flour, one-half cup raisins, a little nutmeg, a little cinnamon, a little allspice, one teaspoonful baking powder, one teaspoonful butter, three tablespoonfuls molasses, a little salt. Bake four hours.

PARKER HOUSE ROLLS.

One quart skimmed milk, one egg, butter the size of an egg, flour sufficient to make a smooth dough, one-half cake of yeast. Raise five hours. Bake three-quarters of an hour at full heat of oven.

OMELET.

Three eggs, one-half cup of milk, two teaspoonfuls of cracker crumbs, a little salt, small piece of butter. Beat the whites to a froth. and beat the yolks up with a fork. Heat a pan in the oven and then butter it. Put the omelet in the pan and cover it. Cook fifteen minutes.

VEAL CUTLETS.

Cut the veal in round pieces, dip them into fine bread crumbs and sautér them a light brown in a little pork. Then put them into a saucepan with a piece of raw veal at the bottom, one-half pint of water, the rind of one lemon, half a carrot, one-half a turnip, one-half an onion cut very fine. A teaspoonful of walnut catsup, a little salt, a shake of cayenne and one teaspoonful of capers. Simmer slowly three and one-half hours or more. When dished chop the yolk of a hard-boiled egg very fine, and sprinkle over with capers. Strain the gravy and pour over.

In the Aladdin oven in order to "sautér" remove the lower tight shelf, place a metal pan on the bottom; lay in the pork and veal, add a little lard, butter or olive oil; when browned on one side turn over. A little practice will fix the time. When first practicing, the oven doors may be kept open. This process when done in a spider on a stove is commonly called frying and is, when done at a high heat, as bad a method as could be devised. When well done at a suitable heat the product is most appetizing.

In this connection I will again point out that in very many recipes which are to be found in the cookery books, directions are given, first to fry some onions or something else to a light-brown, and then to go on using this material in combination with others in the final process of cooking.

In many cases the writers do not mean frying by immersion in very hot fat, but what they really mean is sautéing or frying in a spider, in a thin film of fat, olive oil or butter. Therefore, if any one wishes to try such recipes taken from the books, proceed in the manner laid down above.

Cook onions, meat, apples, sippets of toast for soup, etc., etc., in a pan on the bottom of the oven with a little fat of any suitable kind—butter, oil, beef dripping; and if you must you may use lard, I wouldn't.

SHREWSBURY CAKES.

One cup of flour, one cup of sugar, half a cup of butter, two eggs, a little mace. Beat together and spread as before directed for thin cake.

BOX PUDDINGS.

Slow cooking in vessels which are fitted rather closely with covers so as not to permit evaporation, yields the finest flavor, especially in respect to fruit. A recipe which proved to be very delicious when cooked in the Workman's Pail in one box, was made in the following proportions:

Half a pint of milk, two tablespoonfuls of sugar, one egg, a little pineapple syrup flavoring, two apples pared with the core taken out. Other flavors may be used if preferred.

Hardly any combination of fruit, marmalade, jam with other ingredients cooked in this way ever fails to yield very delicious results.

ANOTHER BOX PUDDING.

Half a cup of crumbs of stale bread, one cup of milk, two spoonfuls of sugar, half a cup of stoneless raisins, one egg, pineapple or lemon flavoring. Cook three or four hours in a pail in tight boxes. Serve with or without cold sauce.

BREAD AND RAISIN CAKE.

Mix in a bowl one quart of white flour with a little less than one pint of water; one-half level spoonful of salt; one-half teacupful of sugar or more according to taste; one-half teacupful of butter softened a little by heat so that it will mix well, but not melted; grate in half a nutmeg, or add a little cinnamon or allspice; add a full teacupful of stoneless raisins or of dried currants, or of chopped dried apples. Mix one third of a cake of Fleischmann's yeast with warm water, add that, stir the whole thoroughly together into a stiff batter. Be careful not to make the batter too thin by using too much water. Raise about three hours, and bake about three hours. In order to get the most thorough bake, and to make the cake as tender as it can be, after it is thoroughly baked and well browned set it on the upper shelf of the oven, place a pan of warm water underneath, lower the flame of the lamp to the lowest point, and keep the cake in the oven for several hours.

The foregoing specific directions may suffice by analogy to enable any one who possesses the required one part of gumption to apply the simple principles which have been laid down, to other specific recipes.

In dealing with the materials of which our food consists, it may be remarked that they may be divided in a broad and general way into certain classes: First, meats; second, fish; third, grains; and fourth, vegetables.

Reversing the order, roots and tubers may preferably be boiled. They are not injured by it; the expansion of the steam in the cells renders the potato mealy and more palatable, while boiling does not injure beets, carrots or parsnips, because there is enough flavor and to spare, in either. Green vegetables, like asparagus, pease and beans, to which squash may be added, are better and of finer flavor when simmered, either in their own juice or with a little water added, than if they are boiled.

All the grains whether converted into the form of bread, or whether cooked in the form of oatmeal, mush, cracked wheat, etc., retain their flavor and are much better, if cooked long at a very moderate degree of heat. Those which are to be eaten in the form of porridge or mush, are much better when cooked for a long time at less than the boiling point, than when cooked in any other way.

In respect to meats, it may be remarked that to boil meat is to spoil meat. While "meat may be boiled to rags," as it is sometimes put, the boiling toughens each separate fibre, while the process takes all the flavor and juice of the meat into the water. All other methods, roasting, imitating a broil, etc., applied to meat, should be done at such a moderate degree of heat as will not distil or dissociate the fats, and not at the high heat of the common stove.

The chafing dish or blazer is a valuable auxiliary to the oven in making sauces or preparing cold meat upon a table. A very excellent recipe for serving cold mutton or venison in a delicious hot form is as follows :

Put the gravy into a chafing dish or blazer, add a saltspoonful of Nepaul pepper, a suspicion of cayenne pepper, a wine glass of sherry or white wine, a dessertspoonful of walnut catsup, two teaspoonfuls or a little more according to taste, of currant jelly. Stir while the ingredients are mixing. Put in the cold meat and warm it, or cook it a little if it is very rare. In dealing with cold beef omit the jelly.

ALL SORTS OF THINGS.

Take any receipt out of any book. Bear in mind that beating up materials or mixing them together, is a different matter from stirring them when they are upon the stove. The stirring process over the fire, which leads to so much discomfort, is merely a corrective of the bad methods of hot iron-stove cooking. It simply keeps the mixture in the pan from scorching or burning under the excessive heat. Stirring upon a stove is in fact a process of cooling which may be wholly dispensed with when the same materials are subjected to a moderate heat in the Aladdin Oven. Acting upon this hint, almost every recipe that is given in any of the books, can be worked out successfully with the Aladdin Oven or in the chafing dish, much better than it can be upon the hot iron stove or range.

FRYING.

I have not attempted true frying in the Aladdin Oven, although I doubt not that if a powerful lamp were used and the frying pan were placed on the bottom of the oven it could be done, but it were better done on an open stove.

The abominable process commonly called frying as practiced in this country, consists in a very bad method of sautéing and is not frying at all. What true frying is is most fully explained by Sir Henry Thompson, M. D., in his most valuable treatise on Food and Feeding, pp. 86, 87, 88, 89, which I venture to copy :

"The process of frying is rarely understood, and is generally very imperfectly practised by the ordinary English cook. The products of our frying pan are often greasy, and, therefore, for many persons indigestible, the shallow form of the pan being unsuited for the process of cooking at a high temperature in oil, that is, at a heat of about 360 degrees to 390 degrees Fahr., that of boiling water being 212 degrees. This high temperature produces results, which are equivalent, indeed, to quick roasting, when the article to be cooked is immersed in the nearly boiling fat. Frying, as generally conducted, is rather a combination of broiling and toasting or scorching ; and the use of the deep pan of heated oil or dripping, which is essential to the right performance of the process, and especially in order to prevent greasiness, is a rare exception, and not the rule in ordinary kitchens. A few words of explanation are necessary in relation to the temperature of the fat which forms the frying bath, a matter of importance to ensure satisfactory results. When a bath of melting fat is placed on the fire and the temperature has risen to 212 degrees, some bubbles come to the surface with a hissing sound ; these are due to a small portion of water, which being converted into steam, rise until all is got rid of. This is not the boiling of the fat, which is now tranquil, and when the temperature has advanced much higher, to something like 340 degrees, a slight vapour is given off. If the fat is permitted to become much hotter, smoke appears, indicating a degree of heat to be avoided, and that the fat has reached what is called the boiling point; when it decomposes and spoils. Before this is reached, the heat should be tested by putting in a slip of bread, which if browned in a few seconds, a sufficient temperature has been attained, and the bath is ready for use. The above remarks apply equally to the temperature of any oil used for the same purpose. The principle on which success depends is, that at the moment of contact with the almost boiling fat or oil, a thin film of every part of the surface of the fish or other object to be fried is coagulated, so that the juices with their flavors, etc., are at once locked up within, and nothing can escape. The bath should, therefore, contain quantity sufficient, and also be hot enough, to effect this result in an instant, after which, and during the few seconds or minutes requisite to cook the interior, the heat is often slightly lowered with advantage. The fish or other material employed emerges, when done, with a surface to which a little oil adheres, but this will drain off, owing to its extreme fluidity when hot, if left on a napkin slanting a minute or two before the fire ; better still on white blotting paper ; and thus it may be served absolutely free from grease. The film of egg often applied to the surface of an object to be fried, is in the same manner instantly coagulated and forms an impermeable case ; while the fine bread crumbs adhering to it take a fine yellow color, being slightly charred or toasted by the high temperature they are exposed to. In order to be free from grease the bread or biscuit crumbs should be very fine, adhering by means of a thin layer of egg previously applied by the brush. If they are coarse and too abundantly used, grease will adhere to the surface or be absorbed by it.

"Excellent and fresh olive oil, which need not be so perfect in tint and flavor as the choicest kinds reserved for the salad bowl, is the best available form of fat for frying, and is sold at a moderate price by the gallon for this purpose at the best Italian warehouses. Nothing, perhaps, is better than well-clarified beef dripping, such as is produced, often abundantly, in every English kitchen; but the time-honored traditions of our perquisite system enable any English cook to sell this for herself, at small price, to a little trader round the corner, while she buys, at her employer's cost, a quantity of pork lard for frying material, at double the price obtained for the dripping. Unfortunately, however, lard is the worst menstruum for the purpose, the most difficult to work in so as to free the matters fried in it from grease; and we might be glad to buy back our own dripping from the aforesaid little trader, at a profit to him of cent. per cent., if only the purchase could be diplomatically negotiated. But so sweet is acquisition by way of perquisite, that none of the present race of cooks appear disposed to part with this particular one for any consideration which can be offered. They are, doubtless, after their fashion, true to their order, and regard in the light of sacrilege any interference with these principles and traditions."

The following rules and specific recipes give the results of an experience of about three years in my own family averaging eight to ten in number, the oven in use being the standard oven under the control of an intelligent cook who has greatly aided me in the development of this system.

I think it better to give the rules and recipes which have been established in household practice in this place, to be followed subsequently by others derived from the practice of Miss Daniell and Mrs. Abel, which will be more accurate. Our own practice has been of the experimental sort, and the record will only give some general ideas of what we have done.

SOUPS.

For soups take any bones or scraps of meat and vegetables that are in a sound condition, cover with water and cook over night, with the lamp low. Set aside to cool, remove the fat, and treat the stock according to taste. After a very little experience the best results will be secured.

SHIN-OF-BEEF SOUP.

Four pounds shin of beef, one carrot, one onion, a few whole cloves stuck in the onion. Cook over night with water to cover. Lamp low.

MUTTON CHOPS.

Set chops in a tin or on a drainer on the bottom shelf of the oven. Sprinkle with pepper and salt. Cook half an hour. Full lamp power.

BEEFSTEAK—TWO POUNDS.

Set in tin on either shelf—the bottom best. Sprinkle with pepper and salt. Cook half an hour. Full lamp power.

ROAST BEEF AND POTATOES.

The oven being already heated, put the sirloin in, timed according to size. One and a half hours before the beef is to be served, having peeled potatoes of moderate size, put them into the pan around the beef ; if of large size, divide lengthwise.

VEAL.

Four pounds of veal, two slices of salt pork. Put pork on top of veal. Baste with a little flour and salt. Cook two hours on lower shelf. Full lamp power.

Cook beef, mutton and lamb the same way, omitting the pork.

CORNED BEEF.

Put six pounds of corned beef in a porcelain kettle. Cover with water. Put cover on and cook five hours on lower shelf. Lamp low.

CHICKEN.

Cut chicken to broil. Take off skin and lay in a dish. Cover with bread crumbs and small pieces of butter. Sprinkle a little pepper and salt. Cook two hours on bottom shelf. Full lamp power.

BAKED CHICKEN.

Cut chicken in pieces and skin it. Put in a baking dish and cover with cream sauce ; cover with bread crumbs and a little piece of butter. Season with pepper and salt. Cook two chickens two and one-half hours.

CREAM SAUCE.

One tablespoonful of flour and one of butter ; one pint of milk.

PORK SMOTHERED IN APPLES.

Cut all the fat off the pork chops and lay in a braising dish ; take one and one-half cups meat stock and thicken with a tablespoonful browned flour and a small piece of butter. Cover and cook one and one-half hours.

GROUSE.

Two grouse roast half to one hour, lower shelf.

GRAVY—One-half pint of milk put in a saucepan over teakettle, two tablespoons bread crumbs, a little onion juice, and pepper and salt to season. Pour over grouse when done.

BEEF.

Cook one hour. Take three pounds solid beef and pour the following gravy over it :

GRAVY—One tablespoonful of browned flour, one and one-half cups French canned mushrooms, one-half cup boiling water.

QUAIL.

Cook one and one-half hours. Make sauce the same as for the grouse and pour over the quail when they are done.

SPICED BEEF.

Put into a large crock a strong pickle—two-thirds table salt, one-third saltpetre—add two great spoonfuls of brown sugar, one flat tablespoonful of allspice, one flat teaspoonful of cloves and one of mace, a dozen pickled onions, half a teacupful of juniper berries, pepper to taste. In this liquor put ten to fifteen pounds, without bone, of the bottom of the round of beef; leave it in five or six days; draw off the liquor, put in a little hot water, place in the oven with cover on, and cook five or six hours.

The spices can be varied, juniper berries omitted, tarragon vinegar added, or any other variation. To be eaten cold. Keep the beef wholly in the pickle, with a clean paving stone on the top.

HALIBUT A LA CRÈME.

Three pounds of halibut.

SAUCE—One pint of milk, one teaspoonful of onion juice, butter half the size of an egg. Season with pepper and salt. Break the fish into small pieces, put a layer of fish in the dish, then pour some sauce over it; then put another layer of fish, and so on till the dish is full. Sprinkle the top with bread crumbs. Cook one and one-quarter hours. High wick, either shelf.

Baked macaroni, one and one-quarter hours, either shelf, high wick. Scalloped tomato, one and one-quarter hours, either shelf, high wick. Baked potatoes, two hours.

STEWED CELERY

Take two heads of celery and cover with a sauce; cook on the bottom shelf from one to two hours. Full lamp power.

SAUCE—One pint of milk, thickened with flour, small piece of butter. Pour this over the celery and sprinkle with bread crumbs.

OYSTER PLANT.

Take two oyster plants and cook in the same way as the celery.

ASPARAGUS.

Two bunches of asparagus; wash and cut off the roots; toast some bread, and butter the slices; put the toast in a dish and the asparagus on the toast; put little piece of butter on the asparagus and sprinkle with salt. Cover and cook two hours on either shelf. Full lamp power.

APPLE PUDDING.

CRUST—Almost one pint of flour, one and one-half teaspoonfuls of baking powder, butter one-half the size of an egg, one cup of milk. Roll thin like pie crust.

Slice five apples and season with a little grated nutmeg and two tablespoonfuls of sugar; put in a dish and cover with crust. Bake one and one-half hours, lower shelf is best. Full lamp power.

CRACKER PUDDING.

Take a quart dish and put in it a layer of common crackers. On the crackers put a layer of raisins, and so on till the dish is full. Take three eggs and beat them up; add one-half pint of milk and vanilla to flavor. Pour over the pudding. Cook one hour, either shelf. May add sugar or serve with sauce.

POP OVERS.

Five eggs, one cup of flour, one quart of milk, a little salt; beat eggs very light, mix other ingredients with egg. Put in little tins or cups and bake one hour, lower shelf. Full lamp power.

COTTAGE PUDDING.

One cup of butter, one cup of sugar, two cups of flour, one cup of milk, two eggs, one and one-half teaspoonfuls of baking powder. Beat up eggs with sugar and butter, add the other things and bake one hour on either shelf.

BROWN PUDDING.

Three cups of flour, one cup of raisins, one-half cup of suet, one and one-half teaspoonfuls of baking powder, one cup of molasses, one and one-half cups of milk, little salt, one-half teaspoonful grated nutmeg, one-half teaspoonful allspice, one-half teaspoonful cloves. Mix and bake three hours.

INDIAN PUDDING.

Two quarts of milk, one cup of Indian meal, one cup of molasses, small piece of butter. Mix and bake four hours with a low wick—either shelf, bottom best.

BIRD'S-NEST PUDDING.

Eight apples pared and cored, one pint of milk, two tablespoonfuls of sugar, one teaspoonful of lemon juice, little salt, grated peel of one lemon. Put the apple in dish, mix the other ingredients and pour over the apples. Bake one hour.

BATTER PUDDING.

Five eggs, one quart of milk, a little salt; beat up eggs; dissolve five tablespoonfuls of flour in the milk, mix and add one-half teaspoonful of vanilla. Bake three-quarters of an hour on either shelf.

GROUND RICE PUDDING.

One quart of milk, heated; stir in one-half cup ground rice; cook on range fifteen minutes; let it cool; beat three eggs very light, mix with the rice and milk; add a small tablespoonful of butter and the juice of half a lemon. Bake one hour.

POOR MAN'S PUDDING.

Two quarts of milk, one tablespoonful of sugar, one-half cup of rice, a little salt. Bake four hours with a low wick.

HASTY PUDDING.

One and one-half cups of yellow corn-meal, two quarts of water; cover. Cook four hours on second shelf with the wick high, or all night with the wick low.

To Fry.—Put a pan on the lower shelf of the oven and put butter in it. Cut the pudding in thin slices and fry in the pan.

SPONGE CAKE.

Six eggs, one-half cup of boiling water, two cups of flour; beat whites of eggs stiff; beat up yolks with two cups of sugar. Mix other ingredients in and bake three-quarters of an hour on lower shelf.

SPICE CAKE.

One cup of butter, two cups of sugar, one cup of milk, four eggs, four cups of flour, one-half teaspoonful ground nutmeg, one-half teaspoonful cloves, one-half teaspoonful allspice, less than one-half teaspoonful of cinnamon, one cup of raisins; beat up eggs. Mix and bake one hour on bottom shelf.

GINGERBREAD.

One cup of molasses, one-half cup of boiling water, one teaspoonful of butter, one and one half cups of flour, one egg, a little salt; beat up egg; add other ingredients, the flour last. Bake one and one-quarter hours.

RYE BISCUITS.

One and one-half cups rye meal, one and one-half cups of white flour, one and one-half teaspoonfuls of baking powder, one-half teaspoonful of salt, one pint of milk, one egg; beat up egg and mix in other things. Bake one and one-quarter hours, either shelf, full lamp power.

BROWN BREAD.

Two cups of rye meal, two cups of Indian meal, one cup of molasses, one teaspoonful of salt, one and one-half teaspoonfuls of baking powder, about one quart of milk, enough to make a stiff batter. Bake three and one-half hours.

WHITE BREAD.

Four quarts of flour; take an equal quantity of milk and water, enough to make a batter; take a little sugar, salt and butter, stir with a bread-mixer or a spoon for ten minutes. Raise five hours in a bread-raiser. Bake two hours with a high wick, or four hours with the lamp low.

Bake white biscuits one hour and a quarter.

GRAHAM BREAD.

Two quarts of Graham flour, one quart of white flour; one-half milk and one-half water to make a batter; one-half cup of molasses, a little salt; stir the same as the white bread. Raise five hours in bread-raiser and bake two hours, wick high. Biscuits one hour.

CORN BREAD.

One cup of white meal, one cup of flour, one and one-half cups of milk, two tablespoonfuls of sugar, one tablespoonful of melted butter, a little salt. Mix and pour in a pan, having the mixture about one inch thick. Bake one hour.

THIN INDIAN CAKE.

One cup of Indian meal ; hot water to make it stiff enough to spread ; one teaspoonful of sugar and a little butter and salt. Spread very thin on tins and bake one hour.

HOMINY.

One cup of the coarsest hominy, two quarts of water; put the hominy in a pot, pour the water in and season with a little salt. Cover and cook five hours. Low wick.

COFFEE.

Soak a cup of coffee over night in a quart of water. In the morning set in the oven on the bottom shelf and heat it twenty minutes. Strain or clean with white of an egg before heating.

POULTRY AND BIRDS. ALADDIN FASHION. CHICKENS.

Cut up the chickens as if for fricassee. Lay slices of cold toasted bread in the bottom of the baking pan. Place the chicken upon that, fill up the interstices with sausages, sliced apples and potatoes cut into halves lengthwise, then sprinkle with bread crumbs. Put some pats of butter with the bread crumbs. Bake about three hours.

In place of the cold toast use coarse hominy samp, which has been cooked all night slowly, with canned tomato or tomato sauce; or put rice upon the bottom of a pan previously soaked so as to have become soft, and use any seasoning you like, either curry powder, or celery salt, or canned mushrooms, or tomato catsup. Fowls may be dealt with in the same way, but should have been previously cooked so as to have become tender; time, according to their size. Cook in tight boxes and use juices to soften the hominy or the rice, making a gravy.

Another way: Cut chickens into small pieces, making cream sauce of one pint of milk, a piece of butter the size of a small egg, a tablespoonful of flour and a little salt. Put together into a crockery jar with cover and bake three hours; fowls, cook five hours. Vary by adding celery seed, celery salt, mushrooms, or other flavoring.

POTTED CHICKEN, GROUSE, OR OTHER BIRDS.

Cut up into small pieces if the birds are large, otherwise cook birds whole, in a brown sauce made with a little onion, salt and pepper and a little brown flour and some water. Put all into a jar and cook slowly until very tender. When preparing the chicken it is better to sauté or fry it in a pan to a light brown before putting into the jar. Another way is to cut off the wings and legs, make sauce, and cook in a jar; then roast the rest of the bird, rare or well done at will; serve together. The wings and legs of ducks and grouse which are apt to be wasted when the birds have been roasted whole, may be potted with the remainder of the carcass. When the meat used in potting has been previously cooked, the preparation should be cooked in the jar from two hours to two and one-half hours; when cooked raw, four hours.

Many persons will be surprised with the results that can be attained with almost any kind of meat in dealing with it by methods corresponding to the above.

PIE CRUST FOR FOUR TEN-INCH PIES.

Take one quart of flour, one cup and a half either of butter,—or half butter and half lard,—or three-quarters butter and one-quarter lard; combine the dry flour with this material as thoroughly as possible ; then moisten with a very little water; then spread in a sheet about one inch thick, ten to twelve inches wide.

Upon the surface of this sheet spread a half cup of the butter, or of the butter and lard; on the surface dust flour from a dredging box, and roll into a roll of about four inches diameter; then cut this roll crossway into eight pieces—four tops and four bottoms; *turn these pieces upon their sides*, and then roll each out to the size and shape of the plate.

The whole secret of making flaky pie crust consists in turning these sections of the roll upon their sides so that the rolling will be across the section. An examination of the condition of the dough when thus rolled, will show why the crust bakes in a flaky condition.

BOILED DISH.

CABBAGE AND SAUSAGE.—Complete success has been attained in cooking corned beef, salt pork, turnips, beets, and cabbage, and also in cooking cabbage and sausage, a very popular French dish under the name of *saucisse aux choux*, in tight tin vessels, subjected to moderate heat for five hours. No perceptible smell was given off in the room, and the cabbage was cooked in a most tender and appetizing manner.

Liver, Bacon and Cabbage may be cooked in the same way, practically without any smell to which objection could be taken.

A very excellent preparation of liver, the cheaper parts of veal, and other somewhat tasteless kinds of meat, is to put them for a day in a hot summer, or longer in winter, in a strong pickle, made with salt, saltpetre and brown sugar, seasoned to taste, either with spice, peppersauce, onions or any other seasoning, then cook with cabbage as above, either with or without bacon or pork. This makes about as strong a food for the least proportionate amount of money that can be devised. Not suitable for people in very sedentary occupations.

PRESERVING FRUIT WITH AND WITHOUT SUGAR.

During the present summer (1891) I have tried some experiments in saving summer fruits, the results of which may not be fully decided before the first edition of this treatise is issued.

I have dealt with Cherries, Strawberries, Gooseberries, Currants, Blueberries and Pineapple.

METHOD.—I have two square tinned copper vessels, each of half the size of the oven, with handles at the front and sides for easy handling.

One is placed upon the close bottom shelf—*not* directly upon the bottom of the oven. The other is placed upon the middle shelf.

Into each pour cold water about one inch deep.

In glass jars of any suitable kind place the fruit—cherries or currants stemmed, gooseberries, strawberries and blueberries carefully picked. To the gooseberries I added a large quantity of sugar, to the strawberries and cherries a very little, to the currants and blueberries none. I filled up oven with cold water; placed in the tinned copper vessels, with covers on loosely; placed them in the oven, cooked all night with a duplex burner, each wick one and one-half inches wide. The heat did not reach the boiling point, but was at about 200° F. in the morning, with the lamp still burning. Gooseberries overcooked and shrivelled, but the syrup very fine. The fruit had all shrunk a little. I filled each jar oven full with water just off the boiling point and closed at once, placing the jars again in the hot water in the pans and cooling off gradually without cracking a single glass jar.

The pineapple was peeled and cut into small pieces, which I forced into open jars of tumbler shape with a marble pestle, without sugar. The skins and butts were placed in a tureen with some water. The fruit in the jars cooked all night in one oven in the manner previously described. The juice extracted from the skins and butts having been strained was reheated to the boiling point, and the jars filled even with 'his liquid and then closed.

In each case the specific flavor of the fruit seems to be developed to the very highest point. At this date the cherries have been consumed, and all the other varieties seem to be in sound condition and likely to keep indefinitely.

The same processes applied to meats and soup stock did not prove a success. I succeeded, however, perfectly in cooking tough meat and converting it into a most tender condition in the glass jars, but it did not keep after sealing.

If this method of saving summer fruit without sugar should prove a success it may be applied in the tropics to some kinds of fruit that are very perishable and cannot be transported in their natural condition.

The surest course seems to be to ripen the fruit and develop the flavor by subjecting it to a low heat, 140° to 150° during the night, forcing it in the morning to about 200°, and then sealing while at that temperature.

FAT PORK COOKED.

A SUBSTITUTE FOR CODLIVER OIL.

A physician, who has been making experiments in the use of the Aladdin oven with a view to nutrition as a method to cure in disease, has called my attention to certain paragraphs in Dr. Thomas Addis Emmet's treatise upon gynæcology, page 97. This paragraph is as follows :

"An excellent substitute for Cod Liver Oil, and one often better tolerated, is fat pork properly prepared. I direct a thick portion of a rib piece, free from lean, to be selected and allowed to remain in soak for thirty-six hours before being boiled, the water being frequently changed to get rid of the salt. It should be boiled slowly and thoroughly cooked, and while boiling the water must be changed several times by pouring it off and fresh water, nearly boiling, substituted. It is to be eaten cold in the form of a sandwich from stale bread, and both should be cut as thin as possible. It is very nutritious, but it should be given in small quantities until a taste for it has been acquired. It may be made palatable by the addition of a little table salt. May be rubbed up in a mortar.

"I some years ago saved the lives of two of my children, who, on different occasions, were suffering from cholera infantum, by feeding them entirely on the fat of pork prepared in this way, and while nothing else would be retained on their stomachs, not only was it retained but it also had a beneficial effect on the diarrhœa."

I have treated fat pork by washing out the salt according to the instructions, then cooking it very long and very slowly in a tight box, the pork resting upon a drainer. It is reduced to the condition in which it seems to be exactly adapted to the purpose named by Dr. Emmet, and when re-salted and made into a sandwich with two slices of dry toast it is indeed very appetizing.

This substitute for codliver oil may serve a useful purpose in dealing with patients who cannot overcome their repugnance to the oil.

ADDENDA.

Since I first tried to put the Aladdin Oven into the market in a tentative manner, after having spent two or three years and a good deal of money in convincing *myself* that it represented a fact and not a fad, I have had a very interesting experience in partially overcoming inertia, especially the inertia of woman. If I were anything but a man of facts and figures, I might write a very amusing article upon "Great Expectations," as illustrated by the demands of those who are not satisfied when I propose to roast in perfection a twenty-eight-pound sirloin of beef or a thirty-pound saddle of venison,* over a single Pittsburgh lamp of the common type.

In addition to what I promise I have been asked if I could not warm the kitchen, heat all the bath water, and run an incubator so as to hatch the broilers as well as cook them, and do a great deal of other work with the lamp! In fact, when I have declined to undertake anything more than to do pretty much all the cooking for a family of eight or ten people with one lamp, some of my correspondents have responded as if they thought I had raised false expectations by naming the Oven "Aladdin."

About 400 people have risked twenty-five dollars each on the experiment in the purchase of an Aladdin Oven. I have yet to hear of the first real failure, although there have been some difficulties in overcoming the prejudices of the cooks. About every other cook takes to the oven at once, and finds in it great satisfaction and saving of labor. The others object more or less to its adoption in the family and fail in making any satisfactory use of it except upon compulsion.

One very cultivated woman who prefers to do the work of her small family herself, has written to me that "it adds much to the enjoyment of her life to be able to give little dinners or suppers to her few chosen friends, without being obliged to serve *roasted mistress* as the first course."

From the sale of ovens thus far made, I might have already recovered the cost of my experiments, had it not been for the contribution of a number of ovens for

---

*NOTE.—One of my friends in England who dwells in a great ancestral hall, in which the roasting for many generations has been done before an open wood fire, in an open tin kitchen in which a historic turnspit keeps the roast turning round and round all the time until it is done, now roasts his own pheasants, venison and four-year-old grass-fed mutton, in the Aladdin Oven in preference to the old method.

charitable purposes. One of the privileges which I find to be connected with such a common-place invention as this, is that one is expected to give an oven to all the charity hospitals and the like. Yet I have made a profit corresponding to the lesser half, if I may use the expression, of what I have spent, and I am now expending this profit in experiments on the Workman's Cooking Pail or Shop Girl's Oven. I beg pardon, I mean an oven for the use of Sales Ladies.

I feel now assured that I shall be able to perfect the Workman's Cooking Pail with a lamp attached, all in one piece. This pail may be charged with about two pounds of food of two kinds at one time, and is so constructed that it may be easily carried by the bail in the hands of the workman to the place of his work. The little lamp may then be lighted so that a hot dinner will be ready at noon with a hot pot of coffee.

I have already succeeded in making a cylinder oven to be placed upon a tin table, which can be charged with three to five varieties of food, amounting to about eight to ten pounds at one time. In this I have also succeeded in baking good bread. In this pail a breakfast for several people can be cooked during the night. A mid-day meal can be in process of cooking while the sewing women or sales ladies are absent at their work. The supper can be warmed over, and the water boiled for the pot of tea in the evening. Bread can also be baked in the evening.

HEATING WATER FOR HOUSEHOLD USE, ETC.

The oven is intended to be used *for cooking*, but directions have been given for warming water on the top. An apparatus for heating water with a second lamp, or for making tea and coffee, or for washing dishes, can be purchased at a low price at any of the shops where common kerosene oil stoves are kept for sale. Water for circulation throughout a house, must be heated from the water-back of the cooking stove, range, or furnace. If I were to build another house, I should attach the apparatus for heating water to my furnace for winter use, and I should place a small water-heater alongside for summer use, as before stated. The kitchen stove which might then be required for warming the kitchen, would be a very small affair, of which the top could be fitted for frying, for boiling water in the tea-kettle and for some other purposes.

The merit of the Aladdin Oven—if any it possesses—as a substitute for about nine-tenths or more of the work of the range or stove, is:

1st. In the saving of food rather than of fuel.
2nd. In saving excess of heat and half the work.
3d. In saving the fine, natural flavors : food cooked a second or even a third

time being generally as appetizing as when first served, and in some instances even more so.

The Oven may be offered as a very complete substitute for a cooking stove in summer houses when hot water for circulation through the house may not be required.

The best measure of the expenditure of oil which I have been able to make, was during the summer of 1889, at my summer place on Buzzard's Bay. We have used no coal for two or three years. A little wood from my wood-lot was burned a part of the day in the cooking stove for laundry purposes and the like. The cooking and the lighting of the household for one hundred days were both done in 1889 with seven dollars' worth of kerosene oil, bought in parcels of five gallons each, at fifteen cents a gallon. I computed the average number of persons, ten ; number of meals, 3,000 ; cost of fuel per meal, about a quarter of a cent or less. I do not, however, consider the economy of fuel as of any importance compared to the economy in the use of the food-material, and the quality of the result of the application of heat under absolute control, in yielding appetizing, nutritious and wholesome conditions.

CARE OF LAMPS.

The great trouble with many of the kerosene lamps is, that in the effort to make them at a low cost, they are slightly or badly made ; therefore they are apt to get out of order. If the demand for these Ovens should increase, special lamps may be made to meet the special conditions of their use.

Lamps should be kept clean—the burners and wick-holder especially—else they will not develop their full power and the food will not be sufficiently cooked in the allotted time. I may name the Pittsburgh lamp as one which can be easily kept clean and in good order.

There is a new burner and also a new wick about to be put into the market which will probably obviate almost all the trouble and difficulty in the use of the lamp.

A stove lamp or heater to be used as an auxiliary may now be found on sale at the stores of the Central Oil Stove Company in New York, Chicago, and in Boston, at 78 and 80 Washington street, which is made like the lamps for illumination, having a central duct, with a circular wick, assuring a practically perfect combustion. The smallest lamp of this type has a wick about eight inches in circumference ; it is fitted with an iron chimney, upon the top of which a considerable quantity of water can be kept hot or made to boil ; frying by immersion in hot fat can be done upon it, and other work which it is not expedient to put into the Oven.

The Central Oil Stove Company now represents a combination of nearly all the makers of the common kerosene oil and gas stoves. These stoves may be very useful auxiliaries where the main dependence is put upon the Aladdin Oven. They may do their kind of work in their way extremely well. They are, however, subject to the passage of the heat and the products of combustion of the lamps directly into the receptacles where the food is placed, which receptacles are ventilated. They may, therefore, taint, scorch or dry the food, in the same way that it may be affected in the ordinary iron stove, unless the work is closely watched.

The point of distinction in the Aladdin Oven as compared to all others, is that the heat may not be raised by the common lamp which is used with the oven of the Standard size, to any point which will be likely to scorch the food, if ordinary care and common sense are applied in putting the food into the oven. The heat will not then be sufficient to distill the fine flavors and send them off in bad smells, while the oven is not ventilated, except when it is used for boiling, because the humidity derived from the food itself make the very best atmosphere in which it can be suitably cooked.

Much larger stove lamps of the type named may enable me to make Aladdin Ovens of larger sizes, on which experiment I am now engaged. The *Standard* size is, however, quite large enough for the average family.

Any intelligent women who can put into this oven the requisite one part of gumption and one part of food, may nourish themselves adequately, changing the bill of fare every day, at not over one dollar a week for the cost of the food and fuel, including the tea and coffee.

With the aid of Mrs. E. H. Richards, bills of fare for seven days, twenty-one meals, have been computed which contain all the requisite elements of nutrition in due proportion, to cost from one dollar to one dollar and a half per week. These dietaries have been given elsewhere.

In Part II will be found a report made by Mrs. Mary H. Abel, the successful competitor for the Lomb prize of $500, offered by the American Public Health Association, whose treatise may be ordered from N. D. C. Hodges, No. 874 Broadway, New York.

Also a report made by Miss Maria Daniell, an experienced teacher of cooking, whose work was more specifically directed to the invention of recipes and methods of seasoning for box cooking in the oven or pail, which may be adjusted to the period of five hours without requiring any attention in the interval; that being the customary period of a morning's work between the hour of beginning and the dinner hour.

THE ALADDIN KITCHEN.

The ideal cooking laboratory which I contemplate adding to my house on the north side, next the present kitchen, will perhaps be built in the following manner: Excavate for basement six feet below surface, lay drains inside and outside the wall, fill up one foot with broken stone covered with asphalt concrete. Lay stone wall to surface in cement, and four feet above with brick laid in cement, inside course of fire brick. On this wall lay hard pine timbers (vulcanized) five feet on centers, which may be painted at once without danger of dry-rot. Lay floor of three-inch spruce, grooved and splined. Finish ceiling between the timbers with plaster board with skin coat of King's Winslow cement. In this basement place a water furnace for heating the house in winter, a water heater to heat water for circulation in summer and for laundry purposes, and fit up for laundry. One window to be fitted for the escape of flies.

THE KITCHEN OR LABORATORY.

Upon the cellar or basement wall as described build a superstructure of three-inch vulcanized spruce plank, grooved and splined. Cover in with roof of the same, resting on vulcanized timbers five feet on centers, covered outside with shingles laid over mortar. Inside finish: Between the roof timbers lay plaster board one inch thick finished with cement plaster; on the inside of the plank lay on porcelain tiles or lay up vitrified brick; lay floor of tiles laid in cement very slightly pitched toward one corner where there will be a scupper for the escape of water. All corners rounded so that there may be no right angles in the room.

This structure will be constructed on the principle of slow burning construction of heavy timber and solid wooden walls and roof, in the study of which the first conception of the Aladdin Oven arose. Solid wood or thick slabs of incombustible wood pulp are the best available non-heat conductors for use in building cooking laboratories or ovens. By avoiding the ordinary cellular construction which is set up by the masters of the art of combustible architecture we avoid the hidden spaces in which rats, mice, cockroaches and fire have free way, and by using thick wood and shingles laid over mortar we are free from the dampness and chill which are generated by walls of brick or stone. In the thick solid plank roof we avoid the customary attic, which serves as an oven in summer and a refrigerator in winter. We also keep our pepper and mustard pots on the sideboard instead of using them as models for the decoration of a crazy roof in the "Queen Anne Style."

The advantage of this method of construction and finishing is that the cleaning may be done with warm water containing a little kerosene oil forced through a hand pump to every part of the room, then wash down with cold water.

Windows: All of customary type save one. The window on the east to be a fly escape.

THE FLY ESCAPE.

This window consists of two sashes placed in reverse of the customary method, the top sash inside, the inner sash so made of wood nearly flush with the glass and bevelled to an even point so that the flies which always walk upward on a vertical wall or window may walk *out* without obstruction. In order to promote the escape of flies especially in the early morning when the rising sun attracts them, lower the upper sash to nearly the middle of the glass in the lower sash; which should be a single plate—of course there would then be an open space between the two plates; draw down and fasten a dark green curtain to the same point, held so close that the flies cannot get behind the curtain; close the blinds at all the other windows. Flies are early risers and always desire to pay their libations to the morning sun; it is a work of charity to give them an opportunity. They are also very intelligent; they will go to the lighted window pane, walk upwards to the open way between the two sashes and then depart. A few perverse ones may remain, and for them a breakfast of fly paper may be laid in front of the window.

LABORATORY FITTINGS.

On the north side, next the outer door opening into a porch, place a Chase Refrigerator, so arranged that the ice can be washed on a platform outside and put into the ice receptacle from without. The refrigerator will be in the northeast corner, next the fly-escape window. In the southeast corner will be the door opening into the pantry, which is now connected with the present kitchen. In fly time the pantry door will be left open at night so as to open the way for flies to go to the fly escape. At the left of the outer door on the north side will be a large marble table or wide shelf with one marble shelf above. On this table all food will be prepared. It will be very slightly pitched to the rear and toward one end, and will be protected at the back by a concave strip of marble attached with cement. At one end will be a water-tap, at the other a scupper hole of large size, through which all refuse may be washed into an open removable bucket.

At the westerly end will be open shelves of vulcanized hard wood for dishes and other apparatus, beneath and a little above the floor level a platform for metal ware protected with sheet copper. No doors and no cupboards will be placed against the inside or south wall. A small iron heater with four holes for cooking vessels on the top and a place for a hooded grill ventilated into the chimney, to be heated with charcoal or coke if gas is not available. Next, a tin-lined copper boiler with hooded escape for steam above it, to be heated by gas. Next, a Case Bread Raiser. Next, a suitable gas or kerosene stove. Finally as many Aladdin Ovens and Cooking Pails as may be required for use in Laboratory practice. One for roasting meats. One for cooking vegetables, stewing or simmering. One for baking bread.

At this point ends the first lesson in the Science of Nutrition which has been developed in a purely empirical manner by the undersigned,

EDWARD ATKINSON.

Brookline, Massachusetts, U. S. A., August 1891.

PART II.

THE SCIENCE OF NUTRITION.

The scientific part of this work will now be presented. In the first part my own experience has been given. I first recognized the profound relation of the food question to all the processes of industry as the most important element in social science when dealing with it many years since in a very practical way. It became incumbent on me to make life easier for a very large number of factory operatives who were employed in a large but isolated factory which was under my charge, on whom a population of about 2500 depended at a time when we were forced by the high and variable prices of cotton, and by the fluctuations of depreciated paper money to run only four days a week. In this way I was gradually led from theoretic and mainly statistical observations to attempt to give a practical direction to this work.

It may have been observed that at the beginning of my study of cooking it was somewhat difficult to overcome a certain grotesque aspect which was presented by the gray-haired manager of factory and insurance company when converted into a cook, fully equipped with an apron and white cap presented by the children, to whom also it was in some senses grotesque; but when the chuck-end of beef that hadn't been cooked long enough to become tender and appetizing, was served for dinner, it did not seem so funny. The same grotesque aspect became more apparent when my friends whom I invited to dinners or suppers, somewhat unconsciously exposed their incredulity, effusively remarking that they had either eaten very heartily and had spoiled their appetites during the day, or else had wholly omitted to lunch so as to be in possession of such a good appetite as would enable them to devour the victuals.

Yet more amusing and sometimes very grotesque have been the interviews with many most excellent and utterly incredulous ladies, who have come to my office in the apparent expectation of meeting some one who would be a sort of combination of Quack Doctor and Head Steward of a hotel. In such interviews it has sometimes been very difficult to maintain a demeanor of dignified courtesy and attention at the risk of very undue hilarity. I have often been reminded of a remark of

Artemus Ward, or one of his type, that "there is a great deal of human nature in men, likewise in some women."

The time of semi-apology on the part of a man who had thus rashly invaded the domain of woman has passed and it gives me the greatest satisfaction to introduce the truly scientific work of the women who have come to my support.

If any one can name another important line of investigation in which any elderly gentleman of an inquisitive disposition, whose early education in "Amusement considered as a force in Christian Training," (See Four Discourses by Rev. Marvin R. Vincent; Troy, William H. Young, 9 First street, 1867,) had been very much neglected, may get as much actual fun as I have in the development of this oven and in the correspondence and interviews connected with it, it may be an aid to digestion to indicate the line on which some one else may work. I have not yet found the other end of this line on which I am working.

In order that those who will take the pains may have the means to carry on this undertaking, certain tables are subsequently given which may be considered approximately correct, but the student will find them suggestive only until a well-equipped

COOKING LABORATORY

shall be established, in which the positive rules of the Science of Nutrition may be more fully established.

The reports of Mrs. Abel, Miss Daniell and Mrs. Richards are based on most valuable tables, prepared by themselves or by Prof. William O. Atwater to whose courtesy I am indebted for permission to reprint his work. The reader may be referred to the articles in the Century Magazine for a most valuable and complete treatment of matters which lie at the very foundation of this science.

DIETARIES FOR THIRTY DAYS.

In order that the greatest economy consistent with sufficient nutrition may be secured without spending more time in saving a few cents a day than the money is worth, the following tables have been prepared on the basis of the retail prices of food purchased in the Boston market in small quantities in the spring of 1891. In consequence of a very short crop of domestic potatoes and the heavy duty or tax imposed upon Canadian potatoes the price of that most important vegetable was excessive at the date of this compilation.

Retail prices of the tougher portions of meat and of the best kinds of grain and vegetables in Boston in the spring of 1891. Vegetables rated at a higher price than usual on account of a great scarcity of potatoes.

THE SCIENCE OF NUTRITION.

	Per lb.		Per lb.
Beef—average of neck, shin, flank, and some of the better portions, omitting expensive cuts,	6 cents.	Halibut Nape,	5 cents.
		Whole milk,	8 "
		Skimmed milk,	4 "
		A good quality of cheese,	16 "
		A fair quality of butter,	28 to 32 "
Beef suet, selected,	5 to 6 "	Corn meal,	2½ to 3 "
Beef, corned,	6 "	Oatmeal,	4 to 5 "
Beef liver,	6 "	Cracked wheat,	5 "
The cheaper parts of mutton,	6 "	Split pease and dried whole pease,	7 to 9 "
Corned pork,	8 "	White beans,	7 to 8 "
Salt pork, varying in different years, now,	8 "	Rice,	6 to 10 "
		Hominy,	4 "
Smoked ham,	12 "	Potatoes, very scarce,	2½ "
Bacon,	12 "	Turnips,	2½ "
Sausage of good quality,	16 "	Onions, scarce,	5 "
Veal, the cheaper parts,	8 "	Beets,	5 "
The cheaper kinds of fresh fish,	6 to 8 "	Carrots,	2½ to 3 "
		Squash,	3 "
Salt cod,	8 "	Cabbage,	2½ "

What can be bought in Brookline, Mass., at retail prices for half a dollar in July, 1891.

1 pound of flour,	2½ cents.
1 pound shin of beef or mutton flank,	5½ "
¼ pound suet,	1½ "
½ pound salt codfish,	4 "
½ pound oatmeal,	2 "
½ pound hominy,	2 "
½ pound corn meal,	1½ "
1 pound potatoes,	2½ "
¼ pound onions,	1½ "
½ pound beets,	2½ "
½ pound carrots,	2 "
½ pound split pease,	4 "
½ pound butter (fair,)	14 "
½ pound dried currants,	4½ "
8 pounds,	50 "

What can be bought for a quarter of a dollar.

12 oz. flour,	¾ lb.	2	c. per lb.,		.	.	1½	cents.
12 " neck beef,	¾ "	5	"		.	.	3¼	"
2 " beef suet,	⅛ "	6	"		.	.	¾	"
3 " salt cod,	¼ "	8	"		.	.	. 2	"
8 " oatmeal,	½ "	4	"		.	.	. 2	"
8 " hominy,	½ "	4	"		.	.	. 2	"
10 " potatoes,	¾ "	2½	"	}	.	.	2	"
2 " onions,		5	"					
4 " sugar,	¼ "	6	"		.	. .	1½	"
2 " butter,	⅛ "	32	"		.	. .	4	"
1 pint milk,	1 "	8	c. quart,		.	. .	4	"
	5 "						23	"
Tea or coffee 2 cups, or one cup each,							2	"
							25	"

Enough for two women at moderate work.

In making up the subsequent tables, Flour Grain, Roots and other vegetables, sugar and a moderate quantity of butter or suet are considered *Constants* in each table. It is assumed that these will invariably be purchased of sound and good quality.

A sufficient quantity of these constants is assigned to each dietary, containing the right proportions of protein, starch and fat, and a sufficient number of Calories to sustain the life of an adult without the addition of the Variables which yield working power.

The term "Constants" is used in a two-fold significance. This part of the dietary may be constant in the relative quantity of each element named, but these elements will also be substantially constant in price in any given year, varying a little in one year as compared to another mainly in ratio to the abundance or scarcity of wheat or potatoes in each respective year.

On this basis of a sufficient quantity of cereals, fat and vegetable food to sustain the life of an adult is dealt with, to which meat, fish, eggs, other vegetables and flavorings may be added to the end that working power may be developed in proportion to the effort which is required in each pursuit. These elements may be added in variable quantities and prices according to their kind.

The following dietaries have been made up to meet the requirements of persons who are occupied in work requiring a moderate amount of physical exertion or of persons who are engaged in sedentary pursuits who take moderate exercise.

Each dietary consists of the two divisions named Constants and Variables.

THE SCIENCE OF NUTRITION. 113

In No. 1, Meat only is added to the Constants.
No. 2, Meat and substitutes, eggs, beans, pease and milk.
No. 3, Fish and cheese.
No. 4, More grain is added, giving an excess of starch at a lower cost, but higher in Calories, because grains are less digestible.
Nos. 5, 6 and 7 are a little richer and more varied.
No. 8 gives a purely vegetarian diet, higher in Calories because less digestible; and higher in cost because animal food is cheaper in cost than vegetables when consumed at a given standard of nutrition.
Nos. 9, 10 and 11 are rich and varied, but are so assorted that they may be adopted by any family at a very moderate cost.

DIETARIES.

Table showing the computation of the elements of Dietary No. 1. The same method has been applied to all.

CONSTANTS WHICH ENTER INTO THE SUBSEQUENT DIETARIES, NOS. 1 TO 12 INCLUSIVE.

ARTICLE.	POUNDS.	PROTEID.	FAT.	CARBOHYDRATE.	CALORIES.	COST AT BOSTON PRICES, 1891.
Flour,	22	2.64	.44	15.18	36,520	$0.55
Grain,	12	1.68	.84	7.60	19,800	.48
Butter,	2	.02	1.73		7,230	.56
Suet,	2		1.78		7,200	.12
Sugar,	2			1.93	3,600	.10
Potatoes,	10	.20		2.10	4,300	.25
Beets, Carrots, Onions, Squash, Cabbage, Parsnips,	7	.13	.03	.50	1,120	.25
For 30 days,	57	4.67	4.82	27.31	79,770	$2.31
For 1 day,	1.90	.155	.160	.910	2,659	.077

VARIABLES IN TABLE NO. 1. SHOWING METHOD OF ANALYSIS APPLIED TO EACH TABLE.

Beef, neck or shin,	12 (including waste)	2.00	.40		5,200	.72
Mutton, neck,	5	.62	.34		2,476	.30
Bacon,	4	.40	2.80		11,840	.48
Beef liver,	2	.40	.10		1,120	.12
Veal.	1	.19	.03		460	.08
Salt Pork,	1	.03	.78		3,160	.08
For 30 days,	25	3.64	4.45		24,256	1.78
Total,	82	8.31	9.27	27.31	104,026	$4.09
For 1 day,	2.73	.277	.309	.910	3,467.5	.136

The food material contained in the following table of Constants would suffice to sustain the life of an adult without the addition of the *Variables* which are given in this and in each subsequent dietary. The Constants may be named *The Life Ration*. The Variables added may be named *The Work Ration*. The price of flour is given at what it would have cost by the sack or barrel in the spring of 1891. All other prices are for small quantities bought at retail. At the time of correcting proofs (September), the price of flour is somewhat higher, and of potatoes considerably lower—the average of all the elements of the dietary about the same.

DIETARY, No. 1.

CONSTANTS.					CALORIES.
22 pounds	Flour,	at	$0.02½	$0.55	
3 "	Oatmeal,	at	.04	.12	
3 "	Cornmeal,	at	.03	.09	
6 "	Hominy,	at	.04½	.27	
2 "	Butter,	at	.28	.56	
2 "	Suet,	at	.06	.12	
10 "	Potatoes,	at	.02½	.25	
3 "	Cabbages,	at	.03	.09	
2 "	Carrots,	at	.02½	.05	
2 "	Onions,	at	.05½	.11	
2 "	Sugar,	at	.05	.10	
57				$2.31	79,770

THE SCIENCE OF NUTRITION.

VARIABLES.

12 pounds	Beef, neck or shin,	at	.06	.72	
5 "	Neck of Mutton,	at	.06	.30	
4 "	Bacon,	at	.12	.48	
2 "	Beef liver,	at	.06	.12	
1 "	Veal,	at	.08	.08	
1 "	Pork,	at	.08	.08	
—— 25				——$1.78	24,256

82 pounds, for 30 days,		$4.09	104,026
2.73 pounds for 1 day,		.136	3,467.5

Cost per week, 95 2/10 cents.

DIETARY, No. 2.

CONSTANTS.   CALORIES.

22 pounds	Flour,	at	$0.02½	$0.55	
3 "	Oatmeal,	at	.04	.12	
3 "	Cornmeal,	at	.03	.09	
6 "	Hominy,	at	.04½	.27	
2 "	Butter,	at	.28	.56	
2 "	Suet,	at	.06	.12	
10 "	Potatoes,	at	.02½	.25	
3 "	Cabbages,	at	.03	.09	
2 "	Carrots,	at	.02½	.05	
2 "	Onions,	at	.05½	.11	
2 . "	Sugar,	at	.05	.10	
—— 57				——$2.31	79,770

VARIABLES.

10 pounds	Beef or Mutton,	at	$0.06	$0.60	
2 "	1½ doz. Eggs,	at	.18 per doz.	.27	
8 "	Beans and Pease,	at	.07	.56	
15 "	Skimmed Milked,	at	.02	.30	
2 "	Suet,	at	.06	.12	
—— 37				——$1.85	29,925

94 pounds, total for 30 days,		$4 16	109,695
3.1 " " " 1 day,		.139	3,656.5

Cost per week, 97 3/10 cents.

THE SCIENCE OF NUTRITION.

DIETARY, No. 3.

	CONSTANTS.				CALORIES.
22 pounds	Flour,	at	$0.02½	$0.55	
3 "	Oatmeal,	at	.04	.12	
3 "	Cornmeal,	at	.03	.09	
6 "	Hominy,	at	.04½	.27	
2 "	Butter,	at	.28	.56	
2 "	Suet,	at	.06	.12	
10 "	Potatoes,	at	.02½	.25	
3 "	Cabbages,	at	.03	.09	
2 "	Carrots,	at	.02½	.05	
2 "	Onions,	at	.05½	.11	
2 "	Sugar,	at	.05	.10	
—— 57				——$2.31	79,770

	VARIABLES.				
10 pounds	Beef or Mutton,	at	$0.06	$0.60	
2 "	Salt Codfish,	at	.08	.16	
6 "	Fresh Fish,	at	.05	.30	
2 "	Cheese,	at	.16	.32	
2 "	Salt Pork,	at	.08	.16	
2 "	Suet,	at	.06	.12	
—— 24				——$1.66	27,448

| 81 | pounds, total for 30 days, | $3.97 | 107,218 |
| 2.7 | " " " 1 day, | .132 | 3,574.0 |

Cost per week, 93 1/10 cents.

DIETARY, No. 4.

	CONSTANTS.				CALORIES.
22 pounds	Flour,	at	$0.02½	$0.55	
3 "	Oatmeal,	at	.04	.12	
3 "	Cornmeal,	at	.03	.09	
6 "	Hominy,	at	.04½	.27	
2 "	Butter,	at	.28	.56	
2 "	Suet,	at	.06	.12	
10 "	Potatoes,	at	.02½	.25	
3 "	Cabbages,	at	.03	.09	
2 "	Carrots,	at	.02½	.05	
2 "	Onions,	at	.05½	.11	
2 "	Sugar,	at	.05	.10	
—— 57				——$2.31	79,770

THE SCIENCE OF NUTRITION.

VARIABLES.

5 pounds	Lean Beef,	at	$0.06	$0.30		
2 "	Bacon,	at	.12	.24		
1 "	Salt Pork,	at	.08	.08		
5 "	Flour,	at	.02½	.13		
2 "	Rice,	at	.06	.12		
1 "	Barley,	at	.05	.05		
2 "	Rye,	at	.03	.06		
1 "	Lentils,	at	.10	.10		
3 "	Whole Wheat,	at	.04	.12		
1 "	Butter,	at	.28	.28		
— 23				—$1.48	38,885	

80	pounds, total for 30 days,			$3.79	118,655
2.7	" " " 1 day,			.126	3,955

Cost per week, 88 cents.

DIETARY, NO. 5.

CONSTANTS.   CALORIES.

22 pounds	Flour,	at	$0.02½	$0.55		
3 "	Oatmeal,	at	.04	.12		
3 "	Cornmeal,	at	.03	.09		
6 "	Hominy,	at	.04½	.27		
2 "	Butter,	at	.28	.56		
2 "	Suet,	at	.06	.12		
10 "	Potatoes,	at	.02½	.25		
3 "	Cabbages,	at	.03	.09		
2 "	Carrots,	at	.02½	.05		
2 "	Onions,	at	.05½	.11		
2 "	Sugar,	at	.05	.10		
— 57				—$2.31	79,770	

VARIABLES.

6 pounds	shin of Beef,	at	$0.06	$0.36	
2 "	round of Beef,	at	.18	.36	
6 "	neck of Mutton,	at	.06	.36	
2 "	Eggs,	at	.18 doz.	.27	
1 "	Cheese,	at	.16	.16	
30 "	Skimmed Milk,	at	.02	.60	
1 "	White Beans,	at	.07	.07	
1 "	Pease,	at	.07	.07	
4 "	Halibut, nape,	at	.05	.20	
2 "	Haddock,	at	.08	.16	

118   THE SCIENCE OF NUTRITION.

3 pounds	Salt Cod,	at	.08	.24		
1 "	Oleomargarine,	at	.16	.16		
2 "	Macaroni,	at	.15	.30		
1 "	Oatmeal,	at	.04	.04		
2 "	Cornmeal,	at	.03	.06		
1 "	Rice,	at	.06	.06		
1 "	Hominy,	at	.04	.04		
—— 66				—— $3.51	41,051	
123 pounds, total for 30 days,				$5.82	120,821	
4.1 " " " 1 day,				.194	4,027.	

Cost per week, $1.35.

DIETARY, No. 6.

CONSTANTS.					CALORIES.
22 pounds	Flour,	at	$0.02½	$0.55	
3 "	Oatmeal,	at	.04	.12	
3 "	Cornmeal,	at	.03	.09	
6 "	Hominy,	at	.04½	.27	
2 "	Butter,	at	.28	.56	
2 "	Suet,	at	.06	.12	
10 "	Potatoes,	at	.02½	.25	
3 "	Cabbages,	at	.03	.09	
2 "	Carrots,	at	.02½	.05	
2 "	Onions,	at	.05½	.11	
2 "	Sugar,	at	.05	.10	
—— 57				—— $2.31	79,770.

VARIABLES.					
2 pounds	Beef, shin,	at	$0.06	$0.12	
3 "	Beef, round,	at	.18	.54	
1 "	Beef, tripe,	at	.10	.10	
3 "	Calves' hearts,	at	.06	.18	
2 "	Pigs' feet,	at	.05	.10	
2 "	Eggs,	at	.18 doz.	.27	
1 "	Cheese,	at	.16	.16	
15 "	Skimmed Milk,	at	.02	.30	
1½ "	Beans,	at	.07	.11	
1½ "	Pease,	at	.07	.11	
4 "	Fresh Fish,	at	.08	.32	
2 "	Salt Cod,	at	.08	.16	
2 "	Bacon,	at	.12	.24	
1 "	Butter,	at	.28	.28	

THE SCIENCE OF NUTRITION.

2 pounds	Macaroni,	at	.15	.30		
2 "	Oatmeal,	at	.04	.08		
½ "	Rice,	at	.06	.03		
½ "	Hominy,	at	.04	.02		
1 "	Sugar,	at	.05	.05		
—— 47				—— $3.47	34,746	
104	pounds, total for 30 days,			$5.78	114,516	
3.5	" " " 1 day,			.193	3,817.0	
	Cost per week, $1.28.					

Dietary, No. 7.

	CONSTANTS.					CALORIES
22 pounds	Flour,	at	$0.02½	$0.55		
3 "	Oatmeal,	at	.04	.12		
3 "	Cornmeal,	at	.03	.09		
6 "	Hominy,	at	.04½	.27		
2 "	Butter,	at	.28	.56		
2 "	Suet,	at	.06	.12		
10 "	Potatoes,	at	.02½	.25		
3 "	Cabbages,	at	.03	.09		
2 "	Carrots,	at	..02½	.05		
2 "	Onions,	at	.05½	.11		
2 "	Sugar,	at	.05	.10		
—— 57				—— $2.31	79,770	

	VARIABLES.				
5 pounds	Beef, shin,	at	$0.06	$0.30	
4 "	Beef, round,	at	.15	.60	
2 "	Mutton, forequarter,	at	.10	.20	
2 "	Mutton, neck,	at	.06	.12	
2 "	Calves' hearts,	at	.05	.10	
2 "	Eggs,	at	.18 doz.	.27	
1 "	Cheese,	at	.16	.16	
15 "	Skimmed Milk,	at	.02	.30	
1 "	Beans,	at	.07	.07	
½ "	Pease,	at	.07	.04	
2 "	Halibut, nape,	at	.05	.10	
3 "	Haddock,	at	.08	.24	
2 "	Salt Pork,	at	.08	.16	
2 "	Bacon,	at	.12	.24	
1 "	Macaroni,	at	.15	.15	

2 pounds	Cornmeal,		at	.04	.08		
2 "	Hominy,		at	.04	.08		
2 "	Sugar,		at	.05	.10		
—— 50.5					——$3.31		42,312
107.5	pounds, total for 30 days,				$5.62		122,082
3.6	" " " 1 day,				.187		4,069.4

Cost per week, $1.30.

DIETARY, No. 8. (VEGETABLE DIET.)

CONSTANTS.					CALORIES.
22 pounds Flour,	at	$0.02½	$0.55		
3 " Oatmeal,	at	.04	.12		
3 " Cornmeal,	at	.03	.09		
6 " Hominy,	at	.04½	.27		
2 " Butter,	at	.28	.56		
2 " Suet,	at	.06	.12		
10 " Potatoes,	at	.02½	.25		
3 " Cabbages,	at	.03	.09		
2 " Carrots,	at	.02½	.05		
2 " Onions,	at	.05½	.11		
2 " Sugar,	at	.05	.10		
—— 57			——$2.31		79,770
With butter in place of suet, adding 44 cents to cost,			44		
			$2.75		

VARIABLES.			
4 pounds Pease,	at	$0.07	$0.28
4 " Beans,	at	.07	.28
2 " Eggs,	at	.18 doz.	.27
30 " Whole Milk,	at	.03	.90
2 " Lentils,	at	.10	.20
3 " Oatmeal,	at	.04	.12
1 " Cheese,	at	.16	.16
1 " Barley,	at	.05	.05
1 " Buckwheat,	at	.05	.05
2 " Macaroni,	at	.15	.30
3 " Cornmeal,	at	.03	.09
2 " Hominy,	at	.04	.08
1 " Rye Meal,	at	.03	.03
1 " Hulled Corn,	at	.07	.07

THE SCIENCE OF NUTRITION. 121

4 pounds	Cabbage,	at	.03	.12	
2 "	Dried Apples or other fruit,	at	.12	.24	
— 63				—$3.24	53,310
120	pounds, total for 30 days,			$5.99	133,080
4	" " " 1 day,			2.00	4,439.3

Cost per week, $1.40.

DIETARY, No. 9.

CONSTANTS.   **CALORIES.**

Less the 2 lbs. of Suet, since this meat will give fat enough.

22 pounds	Flour,	at	$0.02½	$0.55		
3 "	Oatmeal,	at	.04	.12		
3 "	Cornmeal,	at	.03	.09		
6 "	Hominy,	at	.04½	.27		
2 "	Butter,	at	.28	.56		
2 "	Suet,	at	.06	.12		
10 "	Potatoes,	at	.02½	.25		
3 "	Cabbages,	at	.03	.09		
2 "	Carrots,	at	.02½	.05		
2 "	Onions,	at	.05½	.11		
2 "	Sugar,	at	.05	.10		
— 57				—$2.31—12=$2.19	79,770	

VARIABLES.

2 pounds	Beef, shin,	at	$0.06	$0.12	
5 "	Beef, rump,	at	.18	.90	
5 "	Mutton leg and chops,	at	.20	1.00	
3 "	Fowl,	at	.18	.54	
1 "	Bacon,	at	.12	.12	
1 "	Salt Pork,	at	.08	.08	
2 "	Eggs,	at		.30	
15 "	Whole Milk,	at	.03	.45	
2 "	Macaroni,	at	.15	.30	
1 "	Cheese,	at	.16	.16	
10 "	Potatoes,	at	.02½	.25	
2 "	Tomatoes,	at	.05	.10	
2 "	Turnips,	at	.03	.06	
3 "	Sugar,	at	.05	.15	
1 "	Butter,	at	.28	.28	
— 55				—$4.81	38,429
112	pounds, total for 30 days,			$7.00	138,199
3.7	" " " 1 day,			.233	4,606.6

Cost per week, $1.63.

Dietary, No. 10.

		CONSTANTS.				CALORIES.
22 pounds		Flour,	at	$0.02½	$0.55	
3	"	Oatmeal,	at	.04	.12	
3	"	Cornmeal,	at	.03	.09	
6	"	Hominy,	at	.04½	.27	
2	"	Butter,	at	.28	.56	
2	"	Suet,	at	.06	.12	
10	"	Potatoes,	at	.02½	.25	
3	"	Cabbages,	at	.03	.09	
2	"	Carrots,	at	.02½	.05	
2	"	Onions,	at	.05½	.11	
2	"	Sugar,	at	.05	.10	
57					—$2.31	79,770
		VARIABLES.				
3 pounds		Beef, rump,	at	$0.22	$0.66	
1	"	Beef liver,	at	.10	.10	
2	"	Calves' hearts,	at	.05	.10	
4	"	Mutton, loin or forequarter,	at	.20	.80	
2	"	Tripe,	at	.10	.20	
2	"	Pork Chops,	at	.12½	.25	
1	"	Salt Pork,	at	.08	.08	
2	"	Eggs,	at	.18 doz.	.27	
2	"	Pease,	at	.07	.14	
2	"	Beans,	at	.07	.14	
15	"	Whole Milk,	at	.03	.45	
6	"	Fresh Fish,	at	.12½	.75	
1	"	Rice,	at	.06	.06	
1	"	Tapioca,	at	.09	.09	
½	"	Farina,	at	.06	.03	
1	"	Butter,	at	.28	.28	
3	"	Sugar,	at	.05	.15	
48½					—$4.55	39,088
105½		pounds, total for 30 days,			$6.86	118,858
3.5	"	" " 1 day,			.228	3,961.9

Cost per week, $1.60

Dietary, No. 11.

		CONSTANTS.				CALORIES.
22 pounds		Flour,	at	$0.0 -½	$0.55	
3	"	Oatmeal,	at	.04	.12	
3	"	Cornmeal,	at	.03	.09	

6 pounds	Hominy,	at	.04½	.27		
2 "	Butter,	at	.28	.56		
2 "	Suet,	at	.06	.12		
10 "	Potatoes,	at	.02½	.25		
3 "	Cabbages,	at	.03	.09		
2 "	Carrots,	at	.02½	.05		
2 "	Onions,	at	.05½	.11		
2 "	Sugar,	at	.05	.10		
— 57				—$2.31	79,770	

VARIABLES.

2 pounds	Beef, sirloin,	at	$0.25	$0.50		
3 "	Beef, rump,	at	.18	.54		
3 "	Corned Beef,	at	.12½	.38		
2 "	Ham,	at	.12	.24		
3 "	Fowl,	at	.18	.54		
2½ "	Eggs,	at	.18 doz.	.30		
30 "	Whole Milk,	at	.03	.90		
2 "	Salt Cod,	at	.08	.16		
3 "	Cracked Wheat,	at	.05	.15		
2 "	Cornmeal,	at	.03	.06		
1 "	Macaroni,	at	.15	.15		
1 "	Butter,	at	.28	.28		
3 "	Sugar,	at	.05	.15		
— 57.5				—$4.35	43,480.5	
114.5	pounds, total for 30 days,			$6.66	123,250.5	
3.8	" " " 1 day,			.222	4,108.0	

Cost per week, $1.55.

DIETARY, No. 12.

CONSTANTS.   CALORIES.

22 pounds	Flour,	at	$0.02½	$0.55		
3 "	Oatmeal,	at	.04	.12		
3 "	Cornmeal,	at	.03	.09		
6 "	Hominy,	at	.04½	.27		
2 "	Butter,	at	.28	.56		
2 "	Suet,	at	.06	.12		
10 "	Potatoes,	at	.02½	.25		
3 "	Cabbages,	at	.03	.09		
2 "	Carrots,	at	.02½	.05		
2 "	Onions,	at	.05½	.11		
2 "	Sugar,	at	.05	.10		
— 57				—$2.31	79,770	

VARIABLES.

6 pounds	Beef, sirloin,	at	$0.25	$1.50		
4 "	Leg Mutton,	at	.20	.80		
4 "	Lamb or Veal,	at	.15	.60		
6 "	Fresh Fish,	at	.15	.90		
3 "	Eggs, 2¼ doz.,	at	.24 doz.	.54		
2 "	Butter,	at	.30	.60		
15 "	Whole Milk,	at	.03	.45		
2 "	Beans,	at	.07	.14		
2 "	Pease,	at	.07	.14		
2 "	Rice,	at	.06	.12		
1 "	Tapioca,	at	.09	.09		
3 "	Farina,	at	.03	.09		
6 "	Sugar,	at	.05	.30		
				$6.27		50,320
113	pounds, total for 30 days,			$8.58		130,090
3.70	" " " 1 day,			.286		4,336.3

Cost per week, $2.00.

This Dietary contains a customary but unwholesome quantity of sugar.

RECAPITULATION.

POUNDS AND TENTHS PER DAY.

	PROTEID.	FAT.	CARBOHYDRATES.	CALORIES.	COST.
DIETARY 1,	.277	.309	0.91	3,467	$0.136
" 2,	.311	.250	1.07	3,656	.138
" 3,	.279	.309	0.91	3,544	.133
" 4,	.248	.276	1.29	3,955	.126
" 5,	.371	.250	1.16	4,027	.194
" 6,	.307	.311	1.20	3,817	.192
" 7,	.337	.313	1.11	4,069	.187
" 8,	.330	.243	1.50	4,439	.200
" 9,	.311	.294	1.16	3,940	.233
" 10,	.323	.278	1.17	3,962	.228
" 11,	.328	.291	1.14	4,108	.222
" 12,	.319	.324	1.36	4,336	.286

THE SCIENCE OF NUTRITION.

COST PER WEEK.
Omitting fractions of a cent.

No. 1,	$0.95	No. 7,	$1.30
" 2,	.97	" 8,	1.40
" 3,	.93	" 9,	1.63
" 4,	.88	" 10,	1.60
" 5,	1.35	" 11,	1.50
" 6,	1.28	" 12,	2.00

NOTE.—Nearly all the meat included in these dietaries, consists of the tougher portions, which require a long time for their suitable preparation, whether subjected to a roasting, baking, simmering or stewing process.

It will be observed that the quantities given in the foregoing tables are at the weight of the food-material in the condition in which it is bought and sold—the water added in the process of preparation and cooking would vary with each combination. The cost of salt, spice and other accessories will vary with the taste and condition of each consumer. These computations are intended to cover only the necessary elements of customary nutrition.

NOTE.—Copies of these dietaries have been sent by the writer to members of the International Statistical Society and of the Hygienic Association in Europe, and to many correspondents in this country, with blank cards to be filled out and returned. The purpose is to get the prices of the same or corresponding kinds of food in different places, and then to be able to compute the relative cost of complete nutrition in this and other countries. The results of this investigation may be given in subsequent editions of this treatise.—E. A.

The following ration is computed by Prof. A. W. Church (Vide Food, South Kensington Museum Science Handbook, 2d Edition, London, Chapman & Hall, 1889), as being sufficient for an adult man of customary height, five feet eight inches, weighing 154 pounds, who is engaged in work which provides sufficient exercise :

1.	Bread,	18	ounces.				
2.	Butter,	1	"				
3.	Milk,	4	"				
4.	Bacon,	2	"				
5.	Potatoes,	8	"				
6.	Cabbage,	6	"				
7.	Cheese,	3½	"				
8.	Sugar,	1	"				
9.	Salt,	¾	"				
10.	Water alone, or			2 pounds 10¼ ounces.			
	in tea, coffee or beer,	66¼	"	4	"	2¼	"
	Totals,			6		12½	

If the bread be made at home in the Aladdin Oven, it will cost two and one-half cents a pound. If other articles are purchased at Boston prices in July, 1891, the price of this ration will be as follows :

1. Bread,	18	ounces	@ 2½	cents per pound,			2.88
2. Butter,	1	"	@ 32	"	"	"	2.00
3. Milk,	4	"	@ 8	"	"	quart,	1.00
4. Bacon,	2	"	@ 12	"	"	pound,	1.50
5. Potatoes,	8	"	@ 2½	"	"	"	1 25
6. Cabbage,	6	"	@ 2½	"	"	"	.95
7. Cheese,	3½	"	@ 16	"	"	"	3.50
8. Sugar,	1	"	@ 6	"	"	"	.37
9. Salt,	¼	"	@ say				.05
	2 lbs. 10¼ oz.					Cents,	13.50

If the bread were bought at London prices, where home-made bread is almost unknown, but where bakers' bread is much cheaper than it is in Boston, the cost of this ration, which is without meat except the small quantum of bacon, would come to what fifteen cents would buy here. It would be interesting to price this ration in London and in other cities of Europe and also in other cities of this country.

In order that the unlearned may know what the chemistry of the oven and of the digestive organs do with this material, reference may be made to Prof. Church's most interesting and scientific, and, at the same time, very practical treatise.

The subsequent recipes have been tested in a cooking pail which is not yet perfected but may soon be made for sale. The same results may, however, be attained by placing the materials in porcelain or earthen pots or jars with covers—placing them in the Aladdin Oven over a moderate lamp for the time prescribed, or if the cook can be trusted place the cooking boxes on the back of the stove in such a place that the contents cannot be subjected to a high heat.

It will be observed that the time named in most of these recipes is five hours. According to my own experience a great many compounds which are cooked in two or three hours may remain in the oven in closed vessels for five hours or even more without the least injury.

E. A.

SCIENCE OF NUTRITION.

REPORTS.

Miss Maria Daniell, an experienced teacher of cooking, of Boston, has been especially charged with the work of developing methods of five-hour cooking, with a view to practice by those who can give no attention to the process while absent or occupied in other work. Miss Daniell has also computed twelve dietaries for thirty days, in due proportion, which have been verified by Mrs. Richards, and are given elsewhere.

REPORT OF MISS MARIA DANIELL.

When I was first asked to undertake the work of experimenting with the workman's dinner pail, and to make up dishes that might be cooked five hours without detriment, I felt that my field would be a very small one, especially as I must use only the very cheapest kinds of food. I accordingly began with shin and shoulder of beef, and neck and flank of mutton, and by making for the former a nice brown sauce, using the marrow in place of butter, and such seasoning as Worcestershire sauce and tomato catsup, I succeeded in making a very nice dish, which I placed before the members of my family for breakfast, and which they pronounced delicious. For the mutton I made a white sauce, this time using butter and only pepper and salt as seasoning, with a small piece of onion. This, also, was very much liked. But now came the question, would not the people who would be most likely to use the pail oven, think it too much work to make the sauces, and would they, many of them, have the time? It was evident to my mind that I must try putting the meat, flour and seasoning, all into the little crock together, cover it with boiling water, and let the sauce take care of itself. This was so contrary to all established rules of cookery that I did it with some doubt, but I must confess I was surprised at the result. The butter or dripping came to the top, as is always the case in long cooking, but when that was removed with a spoon, a most appetizing dish was presented. In the pail oven having the four triangular dishes and one round one, a whole dinner was cooked, with the most gratifying result. In one of the small dishes was placed some split pease, which had been soaked over night, with just enough water to make a pease pudding; in a second, some coarse hominy, also soaked over night; in a third, some tomato, bread crumbs and seasoning, (scalloped tomato), and in a fourth, some oatmeal for dessert. In the round dish was placed some flank of mutton, with onion, tomato, and a spoonful of Worcestershire sauce, salt and pepper, and a little butter and flour mixed together and mixed with boiling water; then over the whole was poured boiling water, enough to about half fill the tin. The pail oven had been heated for about an hour. The small tins were placed at the bottom, and the one with the stew on top, and left for about three hours. Then the dishes were changed, the stew put on the bottom, and the little dishes on top, and it was left two hours longer, at the end of which time the dinner was dished. The grains were well cooked, and the stew exceedingly nice, and pronounced by a rather fastidious young man, "good enough for any one." After this I tried every kind of meat and fish that could be bought at a low price, and was surprised to find how many things were within my reach. Besides shin and shoulder of beef, neck and flank of mutton, there are such things as calf's heart, lamb's heart and liver, calf's head, pig's head and feet, tripe, beef skirt, beef flank, beef liver, beef heart, salt fish, fresh haddock, halibut nape, and all the different kinds of grains, bacon sausage, kidneys, skim milk and cheese; nearly, if not all, of which have been considered in my thirty days' dietaries, and recipes for the cooking of which I will now proceed to give. With regard to the size of pail oven, I think the size of the one that I have at my house, in which the dinner I have mentioned was cooked, works to better advantage than the larger ones.

I have always cooked in it with the same lamp that I use with the workman's dinner pail; and the dishes ought always to be changed, when they have been cooking about half the time required for the completion of the work; then everything will come out perfect.

NOTE.—These cylindrical ovens have since been somewhat changed in form, and are now for sale.

E. A.

In my work with the "Aladdin Oven," I have not confined myself to cheap food, as the work done in the pail ovens could any of it be as well-cooked in the Aladdin. So I have proved the oven with the more expensive foods, such as fillet of beef, roast and braised chicken, soups and stews of all kinds, puddings and cake, and in every case the result has been simply perfect. I have never before been able to bake cake or bread in such perfection as I do now in the Aladdin Oven, and with such perfect confidence as to the result; indeed, I feel that I cannot say enough in its praise, and I would not part with the one I have, if I could help it, for more than twice its cost.

The tough pieces of meat are rendered tender and delicious by the slow cooking, and soup made in the oven is superior to any I have ever tasted.

I also value, very highly, the Stanyan Bread Kneader, and the Case Bread Raiser. I think they should be included in the furnishings of every kitchen.

I prefer heavy tin dishes for box cooking of meats and soups in the Aladdin Oven, as the stoneware is apt to become crackled and absorbs the fat, and, after a time, becomes rancid.

For the pail oven I like the agate ware best, though the heavy yellow stoneware, being cheaper, does very well, is much better than the brown ware, which I found too thin and apt to burn on the bottom.

Bread, kneaded fifteen minutes in the Stanyan Bread Kneader, raised three hours in the Case Bread Raiser, then just handled enough to make into loaves, and put back in the Raiser for half an hour, will be found in the most perfect condition to bake, having taken less than four hours to prepare. One of the best of bread makers has said: "The ideal bread-pan should be made four inches wide and four inches deep, and any length you wish—ten inches makes a very good-sized loaf."

Miss Parloa says:

GOOD TESTS OF OVEN HEAT WHEN BAKING.

For sponge cake and pound cake, have heat that will in five minutes turn a piece of white paper light yellow.

For all other kinds of cup cake, use an oven that will in five minutes turn a piece of white paper dark yellow.

For bread and pastry, have an oven that will in five minutes turn a piece of paper dark brown.

I think there should be an oil stove, with at least two covers, for heating water and doing such things as cannot well be done in the oven. Such a stove can be procured at the Central Oil Stove Company, 78 Washington street, Boston, for about six dollars.

For the work of a large family, I would prefer a large oil stove made by the Smith & Anthony Stove Company, on Union street, Boston.

With regard to time of cooking in the Aladdin Oven, I allow about one-third longer than in the oven of the coal stove.

It requires fully two weeks of constant use to get the oven seasoned; then this list can be relied upon, we think, as to time:

I find the time given in Mrs. Sterling's letter agrees so well with my own experiments for the cooking of the various dishes that I have had it copied and added to these papers.

Roast Beef (over 8 lbs.),	25 minutes to lb.	
" " (under 8 lbs.),	20 " "
Hindquarter Spring Lamb,	2¼ hours.
Smothered Chicken,	3 "
Roast Chicken,	3½ "
Chicken Pot-Pie,	3 "
Irish Stew (breast and shoulder),	3 "
Cauliflower (put in boiling water),	1½ "
Asparagus,	1½ "
Sliced raw Potatoes stewed in Milk,	3½ "
(Seasoned, when put in, with salt, pepper and butter, very nice.)	
Baked Apples,	2 "
Stewed Prunes,	2 "
Graham Bread (8 in. x 3½ in.),	4½ "
White Bread,	3½ "
Layer Cake,	1 "
Corn Meal Muffins,	1½ "

MISS DANIELL'S RECIPES.

BEEF, OATMEAL AND TOMATO.

1. Cut one pound of shin of beef into small pieces. Season with pepper and salt. Cut up two sausages into inch pieces, roll in flour and put into an earthen dish. Add one cupful of canned tomato, one-third cupful of oatmeal and one teaspoonful Worcestershire sauce, and cook five hours in Aladdin Pail. Cost, 11 cents. Weight, with water added, two and one-fourth pounds.*

PIGS' FEET A LA VINAIGRETTE.

2. Cut up one pound of pickled pigs' feet in small pieces. Put into an earthen dish. Pour over them one and one-half cupfuls of boiling water. Mix one tablespoonful of butter with one large tablespoonful of flour and add to the water. Salt and pepper to taste, and add one teaspoonful of Worcestershire sauce. Cook two hours in Aladdin Pail.

TRIPE A LA VINAIGRETTE.

3. Take one pound of pickled tripe, cut into small pieces. Put it into earthen dish. Mix one tablespoonful of butter with one tablespoonful of flour and add one cupful of boiling water, one teaspoonful of Worcestershire sauce, and salt and pepper to taste. Pour this over the tripe and cook two hours in the Aladdin Pail.

TRIPE A LA CREME.

4. One pound of fresh tripe cut into small pieces. Sprinkle over it pepper and salt. Mix one tablespoonful of flour, one tablespoonful of butter, a small slice of onion and one

---

* All these preparations may be cooked in covered boxes or dishes in the standard Aladdin Oven, as well as in the pail or cylinder oven.   E. A.

cupful of milk. Pour this over the tripe. Taste to see if salt enough, and cook two hours in Aladdin Pail.

BEEF STEW.

5. Cut one pound of shin of beef into small pieces. Pepper, salt and dredge thickly with flour. Add one small onion cut into slices, three or four slices of carrot, and the same of turnip. One teaspoonful of Worcestershire sauce, one tablespoonful of tomato catsup. Cover with boiling water and cook five hours in Aladdin Pail.

BEEF ROLL.

6. Take ten ounces of lean beef from shoulder or shin. Take two ounces of sausage meat and mix with an equal quantity of stale bread crumbs. Cut the meat into slices one-half inch thick and spread with the sausage. Roll up and tie firmly. Salt, pepper and dredge thickly with flour. Put into an earthen dish with one small slice of onion, one-half teaspoonful of Worcestershire sauce and two teaspoonfuls of tomato catsup. Cover with boiling water and cook five hours in Aladdin Pail. Cost, 10 cents. Weight, one and one-fourth pounds.

BEEF ROLL, NO. 2.

Take ten ounces of lean beef from shoulder or shin. Take two ounces of sausage meat with an equal amount of stale bread crumbs. Cut meat into slices one-half inch thick. Mix crumbs and sausage meat and spread on the beef. Roll up and tie firmly. Try out two ounces of fat salt pork and brown the meat in it. Take from the fat and place in earthen dish. Add to the fat one tablespoonful of flour, and brown ; then add one-half pint of boiling water. Boil five minutes. Season with one-half teaspoonful of salt, one-half teaspoonful of Worcestershire sauce and two teaspoonfuls of tomato catsup. Pour over the meat and cook five hours in Aladdin Pail. Cost, 10 cents. Weight, one pound and six ounces.

BEEF A LA MODE.

7. Take one pound of the shoulder of beef and two slices of bacon. Salt, pepper and flour the meat. Put in earthen dish and add three whole cloves, one-half dozen whole allspice, one slice of onion, one teaspoonful of powdered thyme, two tablespoonfuls of vinegar and one-half pint of hot water. If liked turnips and carrots may be added. Cook five hours in Aladdin Pail.

BEEF SKIRT STEAK.

8. Take one pound of beef skirt. Pepper, salt and dredge with flour. Put in earthen dish, and add just enough water to partly cover. Mix one teaspoonful of butter with one teaspoonful of flour and add to water. Cook two hours in Aladdin Pail.

STEWED SAUSAGE WITH POTATO.

9. Put into earthen dish one-half pound of sausage cut in pieces, with one pound of potatoes cut in thick pieces, peppered and salted, one slice of onion, one teaspoonful of flour mixed with a little water. Cover with one-half pint of hot water and cook two hours in Aladdin Pail.

STEAK PUDDING.

10. One cupful of flour, one-fourth pound of suet chopped fine, one-fourth teaspoon'ul of salt, and cold water to make stiff as for pie crust. Roll out one-half inch thick. Have one pound of beef or mutton from shin or neck, well seasoned with pepper and salt. Put in earthen dish with one cupful of water. Cover with paste and cook five hours in Aladdin Pail.

A SHEEP OR CALF'S HEART.

11. Stuff with sausage and bread crumbs. Season with salt and pepper and dredge with flour. Put it into earthen dish and add one tablespoonful of tomato catsup and one small slice of onion. Cover with boiling water and cook five hours in Aladdin Pail. Cost, 16 cents. Weight, one and one-half pounds.

FRESH HADDOCK IN TOMATO.

12. Take one pound of Haddock, salt, pepper and flour it well. Put it into earthen dish, and add one small slice of onion. Cover with strained tomato and cook two hours in Aladdin Pail.

SALT CODFISH IN MILK.

13. Take one-half pound of salt fish, after soaking over night in water, put it into earthen dish and cover with milk. Add two tablespoonfuls of wheat germ meal and cook five hours in Aladdin Pail. Cost, 6 cents. Weight, three-fourths pound.

MUTTON AND TOMATO.

14. Cut up one pound of neck of mutton into small pieces. Salt and pepper and dredge thickly with flour. Put this into earthen dish and cover with strained tomato. Add one tablespoonful of butter, one teaspoonful of Worcestershire sauce and a small piece of onion, and cook five hours in Aladdin Pail. Cost, 17 cents. Weight, two pounds.

MUTTON STEW.

15. Cut up one pound of flank of mutton into small pieces. Season with pepper and salt and put into earthen dish. Add one small onion cut into slices, three or four slices of carrot and the same of turnip. Mix one tablespoonful of butter with two tablespoonfuls of flour. Mix with warm water and pour over meat. Cover with boiling water, add salt to taste, and cook five hours in Aladdin Pail. Cost, 14 cents. Weight, two pounds.

STEWED BEANS WITH CORNED SHOULDER OF PORK.

16. Soak, over night, two-thirds of a cupful of small white beans. Put one-half pound of corned shoulder into earthen dish, add the soaked beans, one small slice of onion and a pinch of pepper. Cover with water and cook five hours in Aladdin Pail.

PEASE PUDDING.

17. Soak one-half pint of split pease over night. Put into earthen dish. Cover with hot water. Add one teaspoonful of butter, one-half teaspoonful of salt, and cook five hours in Aladdin Pail.

THE SCIENCE OF NUTRITION.

INDIAN PUDDING.

18. Scald one pint of skimmed milk and stir into it, while hot, one-half cupful of Indian meal. Add one-half cupful of molasses, one-half teaspoonful of salt and one-half teaspoonful of ginger or cinnamon. Put in earthen dish and add one-half cupful of cold milk, but do not stir. Cook five hours in Aladdin Pail. Butter the dish before turning in the pudding.

PLUM PUDDING WITHOUT EGGS.

19. Butter earthen dish. Take one-fourth cupful of bread crumbs, one-half cupful of flour, one-fourth cupful of suet (chopped fine), one-fourth cupful of raisins, one-fourth cupful of molasses, one-fourth cupful of sweet milk, one-half teaspoonful of soda, one-fourth teaspoonful of salt, one-fourth teaspoonful of cloves and one-fourth teaspoonful of cinnamon. Cook five hours in Aladdin Pail.

BROWN FRICASSEE OF BEEF.

20. Cut one pound shin of beef into small pieces. Dredge with flour, season with salt and pepper, and brown in two ounces of fat salt pork. Remove and add to the fat two table spoonfuls of flour, and brown. Add one pint of boiling water, one tablespoonful of Worcestershire sauce, one tablespoonful of tomato cat. up and one small slice of onion. Pour over the meat which has been placed in an earth·n dish, and cook five hours in Aladdin Pail. Cost, 11 cents. Weight, two pounds two ounces, when ready to cook.

HADDOCK IN TOMATO SAUCE.

21. Melt one tablespoonful of butter in a saucepan. Brown in it one tablespoonful of flour. Add, gradually, one-half pint of cooked strained tomato. Cook five minutes. Add one teaspoonful of salt, one-half saltspoonful of pepper and one small slice of onion. Add one pound of haddock and cook two hours in Aladdin Pail. Cost, 15 cents. Weight, one and one-half pounds.

THANKSGIVING PLUM PUDDING.

22. Boil one pound of raisins in two quarts of milk. Take out the raisins and add to the boiled milk ten eggs well beaten, considerable sugar, and a little salt. Season with spices or vanilla. Cut a stale brick-loaf of baker's bread into thin slices, butter and soak them in the custard. Butter a deep pudding pan and put in alternate layers of the soaked bread and the raisins till the pan is full. Let this stand over night and in the morning fill up with milk. Bake two and one-half hours in a slow oven. Eat with cold sauce.

OATMEAL PUDDING.

23. One and one-half cupfuls of cold-boiled oatmeal. Add one cupful of sliced apples, or one-half cupful of seedless raisins, and one-half teaspoonful of salt. Cook five hours in Aladdin Pail and eat with sugar and milk, or butter, for sauce.

In cooking the grains I used a little less water than is needed when they are cooked at a higher degree of heat, as the evaporation is less in the pail oven. To all grains, except hominy, I allow three times their bulk of water. To hominy, fine or coarse, twice its bulk of water.

BREAD.

One pint of milk and one pint of water (lukewarm), with two tablespoonfuls of butter or dripping melted in it. One yeast cake dissolved in part of this wetting, and flour to make a soft dough, with one tablespoonful of salt sifted into it before adding it to the wetting.

It will take about two and one-half quarts of the best flour. Knead the dough half an hour by hand, or fifteen minutes with Stanyan Kneader, and raise three hours in Case Bread Raiser. Make into loaves, let rise again about one-half hour, and bake in Aladdin Oven one and one-half hours. The time given is for loaves baked in pans four inches wide, four inches deep and ten inches long.

TO MAKE STOCK.

Six pounds of lean beef from the leg, or a knuckle of veal and beef to make six pounds. Cut this in pieces two inches square or less ; do the same with half a pound of lean ham, free from rind or smoky outside, and which has been scalded five minutes.

Put the meat into a two-gallon pot with three medium-sized onions, with two cloves in each, a turnip, a carrot and a small head of celery. Pour over them five quarts of cold water, let it come slowly to the boiling point, when skim, and draw to a spot where it will gently simmer for six hours. This stock as it is will be an excellent foundation for all kinds of clear soups or gravies, with the addition of salt, which must on no account be added for glaze.—(From Catherine Owen's Choice Cookery.)

GLAZE.

To reduce this stock to glaze do as follows : Strain the stock first through a colander, and return meat and vegetables to the pot ; put to them four quarts of hot water and let it boil four hours longer.

The importance of this second boiling, which may at first sight appear useless economy, will be seen if you let the two stocks get cold ; the first will be of delightful flavor, but probably quite liquid ; the last will be flavorless, but if the boiling process has been slow enough it will be a jelly, the second boiling having been necessary to extract the gelatine from the bones, which is indispensable for the formation of glaze.

Strain both these stocks through a scalded cloth. (If they have been allowed to get cool, heat them in order to strain.) Put both stocks together into one large pot, and let it boil as fast as possible, with the cover off, leaving a large spoon in it to prevent it boiling over, also to stir occasionally ; when it is reduced to three pints put it into a small saucepan, and let it boil more slowly. Stir frequently with a wooden spoon until it begins to thicken and has a fine yellowish brown color, which will be when it is reduced to a quart or rather less. At this point watch closely, as it quickly burns. When there is only a pint and a half it will be fit to pour into small cups or jars. It must not be covered until all moisture has evaporated and the glaze shrinks from the sides of the jar. This may take a month.

Of course any strong meat and bone soup can be boiled down in the same way, and where there is meat on hand in danger of spoiling, from sudden change of weather, it can be turned into glaze, and kept indefinitely. I have found glaze five years old as good as the first week.
—(From Catherine Owen's Choice Cookery.)

BEEF A LA MODE.

Take three pounds of fresh beef, trim off the fat ; cut half a pound of bacon into long, slender strips, and lard the beef with it. Mix a few cloves, mace, allspice, peppers, cayenne, a tablespoonful of powdered thyme and two cloves of garlic, with half a pint of malt vinegar. Put the meat into an earthen crock, with a thin slice of bacon under it, add the seasoning and a pint of soup stock, cover the crock, and simmer six hours. When preferred, vegetables may be added, but it is more satisfactory to cook them separately.—(From Thomas J. Murray's Book of Entrées.)

CONSOMME OF FRESH VEGETABLE ROOTS.

Cut in slices two and one-fourth pounds of carrots and the same weight of onions ; put them in a stewpan with some parsley, thyme, shalot and celery, and also one pound two ounces of butter. Try gently to a red color, add eight and three-fourths pints of water, let it boil and skim it ; next put into it a pint and three-quarters of pease and a couple of lettuces ; then add one and one-fourth ounces of salt, one-third ounce of whole pepper, one pinch of nutmeg, three cloves, one and three-fourths pints of dried pease, one and three-fourths pints of white haricots. Let it simmer for three hours in the Aladdin Oven, skim off the grease and strain through a cloth ; then put aside for use.—(Extract from Sir Henry Thompson's Food and Feeding.)

NOTE.—BOX OR PAIL COOKING, OR COOKING IN JARS.

In the course of my experiments in perfecting the workman's pail it has become necessary to adjust the materials and methods to a process of treatment that may be completed in less than five hours, but which may also cover five hours or more without injury to the food. It is assumed that a workman may charge his pail when he leaves home, reach his work at 7 a. m., when he lights the lamp, at 12 m. he is to find his dinner ready; or a woman leaves her room at 8 a. m., returning at 1 p. m. to find her dinner ready.

Nearly all the foregoing recipes have been dealt with in this way in my office dining-room, with most acceptable and appetizing results. Many of the members of the Office Lunch Club give a great preference to the box or pail cooking in closed vessels, as compared to the same food cooked in open vessels in the oven. E. A.

REPORT BY MRS. MARY HINMAN ABEL,

Author of the Treatise on Cooking, for which the Lomb prize of $500 was awarded by the American Public Health Association in 1888. Published by the association.

THE ALADDIN OVEN FOR FAMILY COOKING.

The writer has for six months made use of an Aladdin Oven as an adjunct to a gas stove in her kitchen ; has done with it all the baking and at least three-quarters of the other cooking for the family, the gas stove being used chiefly for roasting, broiling and frying, and where speedy boiling of water was necessary, as for tea and coffee.

As the result of this experience, the following directions and suggestions are offered for the use of those who wish to make trial of the oven.

If we are interested in the advance of this neglected art of cookery, we shall welcome an invention constructed on scientific principles and calling attention to methods of applying heat or the conversion of food materials into food for the stomach. We cannot agree with the inventor of the oven that any method of applying heat will ever make cookery "automatic," but we must own that this part of the process has received far too little attention, the art of cookery having been held to be too nearly synonymous with the art of *mixing*.

The real acceptability of the oven in the average kitchen is only to be settled by trial. It seems to have some drawbacks, as the smallness of the space on which actual boiling and roasting can be done (that is the oven bottom), as this requires some planning for the successful cooking of an entire dinner without help from another stove. But all that is necessary is for the person that buys an oven to take it as the inventor now offers it, as an auxiliary to the range, and then let it make its own way on its merits.* According to the view of those interested in improvements in cooking, this invention ranks high, as it is a distinct advance in the requirement that heat should be fully controllable in amount, and that we should be able to direct it where it is wanted, and prevent it from going where it is not wanted. This latter requirement we may say is met by the non-conducting covering of the stove, and to a degree the controllability of the heat is attained; we could say fully so if the cooking temperature for an oven full of food or for any oven dish could be more quickly reached. This the inventor promises us by the use of a more powerful lamp.

SLOW COOKING. - It is for this that the oven is of the greatest value, and we have by its use an opportunity to prove or disprove what has so long been asserted, that a long application of heat at a temperature lower than has been in general use is necessary to develop the full digestibility and best flavor of many kinds of food, as the coarsely ground grains, the tougher cuts of meat, fruits, dried pease and beans, and also of many dishes like Indian pudding, famous in the days of the old brick oven, but almost unknown in their best estate to this generation. This slow cooking has little chance of trial in the ordinary stove, although

---

* The inventor is now making experiments on larger sizes and more powerful lamps, but he is at present of opinion that two of the present standard ovens will be more convenient and useful than one large one. E. A.

the oven of a gas or gasolene stove has been used successfully by the writer in this way, a low degree of heat being possible and steadily applied from the sides as well as the bottom.

CONSTRUCTION OF THE OVEN.

Every housekeeper knows that a gas or kerosene stove has enough points of difference from the ordinary coal range so that she must give it a little study before she can be uniformly successful in cooking with it. In the same way, the Aladdin Oven has some points peculiar to itself which should be carefully noted to begin with. Notice its construction:

It is a simple iron box, closed in front by a door, and having an opening in the top that communicates with a tube to let off any superfluous steam. This box is surrounded by another whose top and sides are made of non-conducting material for the purpose of holding the heat. A standard, on which this box is set, and a lamp underneath completes the apparatus.

Now it is quite evident that this oven is to cook by heat that is slowly accumulated and then held to its work by the non-conducting character of the oven walls, and it will be of first importance, therefore, to keep it closed as far as possible.

Second, we shall expect the cooking to be done more slowly than in the ordinary stove, at least in the beginning of the process and that more time must be allowed for any given dish.

Third, we shall doubtless find the bottom of the oven hotter than the top, and the different dishes of which the meal is composed must be arranged in the oven accordingly. These limitations must be kept in mind that we may know just what we can reasonably expect of a cooking apparatus constructed as is the Aladdin Oven.

THE LAMP.—It should be also noted that the condition of the lamp is of great importance. It must burn evenly and to its full height, unless purposely turned down for slow cooking. When the lamp is new it seems to take care of itself, and these conditions are met without trouble, but with use the air passages, of course, become clogged and the wick may not push up and down easily. If then taken apart and the working parts boiled in soda it will again be satisfactory.

Although the lamp will burn for eight hours after filling, the flame will decrease in height after the first few hours, and this, if not noticed, will lead to disappointment. We have simply to refill the lamp.

Again, do not continue to use a chimney that has been broken at the top, though the break may not seem to interfere with the draft. The heat of the oven will be best economized if it is so placed that it is not exposed to strong drafts, which must also be avoided as making the lamp smoke.

TEMPERATURE OF OVEN.

The empty oven after being heated for two hours has been found on the average at 360° F. This in the spring, in a kitchen not otherwise heated and the oven placed in a somewhat drafty position between two doors. Under more favorable circumstances the heat was found to be 10° to 20° greater. This point is of importance, since it shows the maximum heat of the oven and the degree that can be communicated in time to the contents.

To learn at what rate the heat of the oven was lost to the surrounding air the lamp was extinguished after the oven was fully heated, and after an hour the reading on the thermometer again taken. It was 150°. How much heat was still held in the walls of the oven was shown by again lighting the lamp and testing at intervals. After fifteen minutes the oven was found to have again reached its average of 360°.

Other records with the thermometer show how the temperature is affected by what is put in the oven to be cooked. Into the oven as heated above were put four cookers, containing six quarts of water and vegetables, at a temperature of 70°. This brought the thermometer down to 150°. In one and one-half hours after it registered 270°, and in another half-hour it had risen to 325°.

A two-pound roast of beef was then put in, causing another fall in the reading, and it was not till one and one-half hours later that 310° was reached. The cookers in this case were covered, allowing very little escape of steam. These and similar observations lead us to a few general rules as to the use of the oven:

1. Since it does not reach its full heat for an hour or somewhat more, that time must be allowed before it is attempted to bake or roast or to cook in any way requiring a high heat at the beginning. Dishes that do not require this high degree of heat may be put in at convenience.

2. The time necessary for cooking any given dish will depend on the size of the dish, where it is placed in the oven (whether top or bottom), and on how much other cooking the oven is expected to do at the same time. The greater the quantity of food to be raised to a certain temperature, the longer it will take.

3. The boiling or the baking temperature once reached, it is evenly held (unless the oven is temporarily cooled by other additions), and the time for cooking a dish may be accurately reckoned from that point, according to rules familiar to us in using other stoves.

Food to be cooked in the Aladdin Oven may, of course, be prepared according to any recipe preferred, only slight changes being necessary to adapt to the oven. The main thing to be remembered is that evaporation is but slight, and little allowance is to be made for "boiling away."

In general it may be concluded from what has been said of the character of the oven that the only dishes that require special attention are those that require a high degree of heat at the beginning, as roast beef, or those that should have a continuous application of heat at the boiling point, as grains and vegetables containing a large proportion of starch. Rice, for instance, may be thoroughly softened at a temperature considerably below the boiling point, but the raw taste reveals the fact that the starch grains have not been fully ruptured. But the oven cooks easily, on both top and bottom shelves, all kinds of meats on any but the roasting principle, eggs in their many combinations, fruits, and most grains and vegetables.

BREAD BAKING.

It will be found most convenient to so time the making of the bread that it will be ready to go into the oven when the dinner comes out. Otherwise, heat the oven for an hour beforehand. Use the oblong or brick loaf pans, the loaves baked in which will weigh about one and one quarter pounds; the oven will accommodate six of these at once. Do not allow the bread to become quite as light as if to be baked in the ordinary oven, where its rising is so soon checked. Place half of the bread on the lower shelf and half on the middle, and exchange them midway in the time of baking. See that the lamp is burning perfectly, as its full heat is required.

TIME.—Bake rolls, rusks, etc., on the lower shelf, with one grate below, and allow three-quarters of an hour, or, if very thin, one-half hour. Bake four loaves of size above described, one and one-half hours. For six loaves allow one-quarter to one-half hour more, decreasing the heat a little at the last.

Stirred bread, which can lose more water without becoming dried, may be baked three hours or longer, and this time is necessary for bread in which other grains are used, as corn and rye.

Unless this baking time is much exceeded, hardness or dryness in the bread is not to be charged to the baking, but rather to an undue proportion of flour in the mixing. The writer has carefully compared bread baked in the Aladdin Oven and bread baked in a gas oven, weighing the loaves before and after baking. The result showed almost exactly the same loss in weight, whether the loaves were baked in the Aladdin Oven one and one-half hours at a temperature that started at 360° F., or in the gas oven three-quarters of an hour at 400° F.

This six months' use of the oven for bread baking may be considered a fair test. It is greatly liked in the writer's family, and the ordinary quickly baked bread is no longer satisfactory.

SODA BISCUITS, GEMS, ETC.

Doughs of this character bake on the bottom with one shelf under in one-half to three-quarters of an hour. Short cakes are baked in the same length of time and afterward split and filled. The thinner or drop mixtures made of wheat, rye, or corn, are baked in three-quarters to one hour in gem pans, which it is better to heat in the oven while the dough is being made. All breads mixed with water have been found to require a somewhat longer application of heat than those mixed with milk.

GRAINS.

In cooking mushes made from the coarsely ground grains, the oven will be found very valuable. The long, slow cooking needed for oatmeal, cracked wheat and corn in all its grades, from flour to hulled corn, is here easily performed without watching, or such an outlay for fuel as amounts to more than the food material is worth.

Mix any of these grains with the same proportion of water, cold or hot, that would be used in the double boiler. Put into a covered cooker and keep on the oven bottom until the boiling point is reached, when it may be removed to the upper shelf to finish cooking. For the "granulated" varieties an hour will be enough, for other kinds two hours or longer.

Various trials have failed to show any difference in the taste of these mushes, whether they are put into hot or cold water. The only requisite seems to be that they be cooked long enough, and at some time in the process be brought to the boiling point.

RICE.—A number of experiments in the cooking of rice in the Aladdin Oven has led to the conclusion that rice in order to be perfectly cooked must be put into boiling water, three parts of water to one of rice, and placed in the already heated oven to remain half an hour, or, if the quantity is large, somewhat longer.

SOUP STOCK.

The cook-books unite in prescribing a "gentle simmer" as the proper thing in making soup stock, but the question how to maintain it for a number of hours on an open range has puzzled many a housekeeper. The water in the pot will too often be found either at a mad boil or entirely still, and the much peeping into the pot to see how matters are progressing dissipates the fine aroma that should be kept in the soup. Even an adjustable flame beneath the pot does not solve the difficulty, as the air is constantly cooling the sides and the top. Only in a steam bath or by the oven heat before referred to can the ideal result be obtained, and only small quantities can be conveniently so made. But the Aladdin Oven seems to meet all the requirements for soup making, and may be considered a great success in this line. It has been used in the New England Kitchen for a year and a half in making daily many gallons of beef broth of the best quality, and it is no less successful in making stock for the family soup.

The meat and bones are put into cold water as usual, and the flavors may be added at the same time, as they will not be dissipated with this method of cooking. It is more convenient to make soup when no other cooking is going on, as over night. If the quantity is large, the flame of the lamp should be kept at full height; if small, turn the flame partly down, or use a smaller lamp. Time, six, eight or ten hours. When the stock has cooled and the fat been removed, it is an easy matter to season and re-heat the needed quantity each day in the oven.

VEGETABLE AND FISH SOUPS.

Requiring less time, and having no fat to be removed, these do not need to be cooked the day beforehand, unless in the case of pease, and beans, that require five hours' cooking, and are improved by still more. The various chowders are all well cooked in the oven. They may be mixed according to any recipe, very little allowance being needed for boiling away.

FISH.

Sir Henry Thompson, in his lectures before the British Fish Commission some years ago, recommended baking as the best method for cooking all varieties of fish, and, since the heat in this case should be moderate, the Aladdin Oven has been found well adapted to this kind of cookery.

The fish may be stuffed or not, but in any case should be well seasoned. Thin slices of pork should be laid over the fish to baste it, or it may be baked in a sauce. It has been found best to bake fish on the platter from which it is to be served, to retain all the juices, and it is better to bake on the middle shelf than on the oven bottom. In an already heated oven one hour will be sufficient unless the fish are very large.

SCOLLOPED, CREAMED AND CURRIED FISH.

These various dishes, made of fish already cooked, are to be mixed in the dishes from which they are to be served, and placed in the oven until thoroughly hot.

MEATS.

ROASTING.—So far the writer has not succeeded in producing a roast of beef that is satisfactory to those who demand the rare and juicy interior ; for this a very high degree of heat at the beginning of the process seems to be necessary. But a tolerable roast may be obtained by placing the piece of meat in a pan already heated and containing a little fat, and putting this directly on the oven bottom.* The upper surface must be spread with butter or fat to keep it from drying, and the meat must be turned over midway in the time. One hour and a half has been found sufficient for a small roast of two to three pounds ; two hours or more must be allowed for a large one. It is best to devote the entire heat of the oven to it and to first heat the pan before putting in the meat, using only enough fat to prevent the meat from sticking.

In roasting veal or pork, where the rare interior is not desired, the result is very satisfactory. Chickens and small birds are apt to become dried by the longer application of heat and less opportunity to baste. To avoid this they may be larded, and if the flavor of pork is disliked, it is the better way to steam them closely covered in a very little water, or to cook them in a well-flavored sauce.†

BRAISING, STEWING, POT ROAST,

are all well adapted to the oven, as also the cooking of corned and smoked meats, requiring, as they all do, a long application of heat somewhat under the boiling point. Reckoning from the time this is reached, it is only necessary to allow the number of hours required in the ordinary stove, but it is well to remember that none of these meats are easily over-cooked. In all cases the cooker should be tightly closed.

RE-COOKED MEATS.

There are a great number of dishes whose basis is meat already cooked, to which other ingredients and flavors are added in a sauce. The success of the dish depends on the slow heating that shall mingle the flavors without really raising the meat to the boiling point (which is also the toughening point for many kinds that have been once cooked), and for such work the heat of the ordinary stove is often too great. These dishes can be cooked on the bottom of the oven, with the tight shelf underneath, in three-quarters of an hour, or better above, allowing twice as much time.

EGGS.

It is now very generally understood that eggs should be cooked considerably below the boiling point to keep the white from becoming horny and indigestible. This temperature is 170° to 180° F. As a little more or less time in the process is of no consequence if this temperature is not passed, a little practice will make the Aladdin Oven of great use in this branch of cookery.

BOILED EGGS.—Put the eggs in cold water and place in the oven until the water reaches 180° F. Or, if inconvenient to test this, let the water first come to the boiling point, a half

---

* Some slight changes made in the oven since these tests were applied have enabled simple rules to be made which have been given by the writer elsewhere, which may give a preference to this oven for roasting over any other appliance. Such has been the experience of very many persons in practice.

† See instructions given elsewhere for box cooking.   E. A.

pint for each egg, put in the eggs and remove the cooker from the oven, letting it stand for five minutes. Or the proportion of water may be less and the cooker be left in the oven five or ten minutes, anyway to communicate the desired degree of heat to the interior of the eggs.

BAKED EGGS.—Break each egg into a buttered cup and set the cups into a pan of water, which place in the oven till the whites become slightly opaque. These can be served on thin slices of ham that have meanwhile been frying in a pan on the oven bottom. Or the eggs may be baked on a stoneware platter previously covered with a layer of heated hash, rice, macaroni, asparagus, or whatever else may be desired. Make dents in this mound with the bowl of a spoon for the reception of each egg.

OMELET.—Prepare as usual and cook for about five minutes in a pan on the bottom of the oven. Fold and serve.

CHEESE DISHES.

The various mixtures of milk, eggs, bread, etc., with grated cheese, as fondue, fondamin, and ramokins, are easily baked in the Aladdin Oven. The time must be suited to the size of each dish.

VEGETABLES.

Young and tender vegetables, spinach, asparagus, pease and beans, corn, tomatoes, summer squash, and half-grown beets should be steamed in a very little water rather than boiled. If the water in the cooker is first heated to the boiling point they will require, on the oven bottom, but little more time than ordinarily allowed. Or they can be put at once into the water to avoid again opening the oven. When they have once reached the boiling point they will finish cooking on the upper shelf in a somewhat longer time.

Winter squash should be baked in its rind until tender, when the inside can be scraped out and seasoned.

Beets, if full grown, will require two to four hours.

Potatoes and macaroni, in which the starch constituent is large, should be kept for some time at the full boiling heat; but it seems to make no difference with the result whether they are put into cold or hot water, or how long they are in coming to the boiling point. It is important, however, that they be taken out and drained when tender. They may then be kept hot in the oven for a considerable time without injury. If put into boiling water and kept on the lower shelf, the time is one-half to three-quarters of an hour, as on the ordinary stove. If then transferred to the upper shelf, more time must be given.

It is important to know how to treat a vegetable so much used as is the potato; by this method the result has always been good, and by no other way of cooking known to the writer will macaroni be so tender and increase so in size while remaining entirely unbroken. Potatoes may also be baked on the oven bottom in one to two and one-half hours, but the writer has not found this method always satisfactory in producing mealy potatoes. If placed directly on the bottom, they must be turned in order that they may bake evenly.

CAKE.

Cake has been successfully baked only on the lower part of the oven, with two shelves interposed between it and the bottom. Patty pans and layer cakes bake in one-half to three-quarters of an hour, cookies in less time, loaves require one and one-half to two hours. If small forms are used for baking the ordinary cake mixtures, and with the full heat of the oven, the result is satisfactory, but loaves are so long in rising that the cake has been found to be coarse-grained as compared with the same mixture baked in a gas oven a shorter time. For fruit cake and others requiring long, slow baking, the oven has been found to be well adapted.

PUDDINGS.

With puddings of all kinds the work of the oven has been very satisfactory; many kinds have been tried and without anything like failure. Especially good have been those puddings having a custard foundation, because the delicate curdling point is not so easily passed as in the ordinary oven; excellent also are all fruit and suet puddings and Indian puddings; for these four to six hours must be allowed on the middle or top shelf. A custard pudding of medium size bakes in an hour.

PIES.

With pie crust, especially puff paste, the results have not been as good, as it seems to require high heat to make it light and reasonably wholesome. Not many experiments, however, were made in this line, a baking powder crust baked as a short cake or rolled as in roly poly pudding being found so much more satisfactory to use with fruit or even with the ordinary lemon pie filling.

FRUIT.

Any fruit, fresh or dried, may be cooked slowly in the oven, the longer the better, it would almost seem, from the surprisingly fine flavor gained by some kinds, as apples and pears when cooked till they turn red. In all cases cook the sugar with the fruit. This effect has suggested that perhaps fruit is ordinarily not cooked enough. A longer application of heat may have a ripening effect, so to speak, that improves digestibility as well as flavor. It is also an advantage that fruit can be thus thoroughly cooked and yet keep its form perfectly. Cranberries, two parts of water and one part sugar to four parts of fruit, cooked for two hours in a covered vessel, are very attractive in appearance and perfect in flavor.

Entire dinners cooked in the oven have required some planning, and a few specimen dinners will show how they may be managed. Oblong cooking vessels will best economize the oven space, but two of the ordinary round cookers, holding enough for a family of six or eight, can be placed on the bottom of the oven and two on the middle shelf for slower cooking, while the spaces between may be used for quart or pint cups, in which fruit, etc., can be cooked, or for patty pans filled with cake, tarts or puddings. The estimated time for cooking is for covered vessels.

DINNER NO. 1.

Tomato Soup. Beef Stew, with Potatoes and Dumplings.
Macaroni. Gingerbread. Stewed Fruit.

The lamp was lighted at nine o'clock for a dinner to be served at one, and at the same time on the middle shelf were placed materials for a tomato soup and the fruit to be stewed, while on the lower shelf was put a cooker, half filled with water, for the macaroni and another containing the beef stew. This dish was prepared as usual, little water being allowed for boiling away, and each piece of meat was floured to give the right consistency to the gravy. The potatoes, being old, were first scalded to remove the strong taste, and they were then laid upon the meat. This stew was put in an iron pot with a close-fitting cover. The oven was then closed until twelve o'clock. When the macaroni was put into the cooker on the lower shelf, in which the water was found to be boiling, and the spaces between the cookers were filled with patty pans of ginger cake. A little before one o'clock the oven was opened to remove the soup, to drain and season the macaroni, which was then returned to the upper shelf to keep warm, and also to drop, in spoonfuls, on the stew the dumplings (a simple baking powder mixture); this was allowed to steam close covered for twelve minutes. This dinner required, besides the putting together of materials, which is the same in any case, only slight attention once between lighting the lamp and beginning to serve the dinner. This dinner was varied by using different meats for the stew, and by substituting for the macaroni any tender vegetables that were in season. Soup stock made the day before was heated on the upper shelf.

DINNER NO. 2.

Pea Soup.   Boiled Tongue.
Boiled Beets.   Winter Squash.
Baked Apples.   Suet Pudding.

As all these dishes required long cooking, they were prepared as far as possible the night before and put all together into the oven at seven or half-past, when the lamp was lighted. The tongue and the pease (already soaked over night) were placed on the bottom with no shelf intervening, the other dishes above. The beets are supposed to be reasonably tender to begin with, not the withered variety of the late winter, which no amount of cooking will make really edible.

This dinner was found very convenient when the entire forenoon was wanted for other work than cooking. It was varied by other dishes requiring the same long cooking, a half hour more or less making no difference in the result. Such dishes were: Boiled ham, hulled corn, hominy and other coarse cereals, pork and beans, bean soup, etc.

It may here be added that this preparation of the dinner as far as possible the night before, though the housekeeper may consider it a great innovation, often changes a hurried morning into a leisurely one and rescues the dinner from failure.

DINNER NO. 3.

In dinners one and two the food was cooked at the temperature of boiling water or that slight degree above it due to the presence of various soluble constituents of food; some of the food was even cooked below the boiling point. But if too much cooking is not attempted at once, the temperature of the oven may be brought to considerably over 300°, and even 300°

has been found to give to roast beef much of its characteristic flavor. The lamp was lighted at nine o'clock for a one o'clock dinner to consist of

        Tapioca Soup.   Roast Beef (2lb. to 3lb.), with Yorkshire Pudding.
        Baked Potatoes.   Stewed Parsnips.
        Cranberries.   Bread Pudding.

The oven was allowed to heat for an hour and then the pudding and cranberries were put in on the oven bottom. At eleven o'clock the potatoes were put in also on the oven bottom, but with a tin under to keep from burning, and the cranberries and pudding were transferred to the upper shelf to finish. At the same time a pan containing some fat was put in to heat for the beef. The beef was put in at 11.30 below, as also the Yorkshire pudding above, in patty pans. The potatoes were turned, and the parsnips in a very little water were put on the lower shelf. The stock was set on the upper shelf to heat, and the tapioca put to cook in a cup. In three-quarters of an hour the meat was turned, the cranberries taken out, the parsnips put on the upper shelf in their place to finish cooking, thus making room below for the Yorkshire puddings, which need some of the bottom heat to make them brown. At one o'clock all was found cooked and ready.

OTHER USES FOR THE OVEN.

AS A WATER HEATER.—In a family where the Aladdin Oven is used for cooking, the necessary heating of water for dish-washing, etc., can be done by filling the oven with pots of water as soon as the dinner is removed. It is better, however, to have a copper-bottomed tin boiler made to fit the entire oven space. Such a one will hold about eight gallons, and will be made by a tinsmith for $1.50. It should have a ring on the front side and one at each end to assist when it is drawn in or out. Have made, also, a wooden bench of the exact height of the oven bottom, and this placed in front of the opened door will make the removal of the boiler full of hot water, safe and easy even for one person. If properly managed, this boiler will heat enough water to do the washing for a small family. It can be heated over night for use early in the morning, and will be found nearly at the boiling point, though the lamp may have gone out some time before. Another boilerful can be heated and removed in time to cook the dinner.

This quantity of hot water will also bring the bath water for one person to the required temperature, and this method is certainly to be preferred in summer to heating the whole house with the kitchen range when the heat is not otherwise needed.

It may be mentioned that it has been found to be an excellent plan in houses heated by a furnace to run a pipe from the water tank through the furnace, obtaining in this way an unlimited supply of hot water, day and night, during the winter. In summer a judicious use of the Aladdin Oven during the hours when it is not needed for cooking will go far toward supplying this want.

AS A FRUIT CANNER.—The boiler above described is to be put to a use that will make it welcomed by any housekeeper in summer. Put a layer of straw in the bottom to prevent "bumping" and pack it full of glass fruit jars previously filled with fruit and hot syrup. It will hold eight of the quart size. Partly fill the boiler with water and let it remain in the oven till its contents have reached the boiling point, when the jars are removed and sealed in the usual manner.*

In the case of corn, pease and other vegetables requiring long cooking or re-cooking at intervals, the advantage is obvious, as the long-continued heat necessary for killing fermentative germs and their spores can here be well maintained.

Housekeepers often have opportunity to buy toward night perishable fruit very cheap, and would oftener do so did it not necessitate sitting up half the night to can the fruit. It would seem that this method could easily be used; it would only be necessary to refill the lamp in the morning and bring the fruit again to the full boiling point before sealing.

The oven may be used in fruit canning without this boiler, the fruit and sugar being put into a porcelain kettle and kept there until thoroughly cooked. As there is no danger of burning, no water need be added. All fruit tried by the writer was perfectly cooked without being broken, only in the case of black raspberries the juice seemed to be too much drawn out of the berries, leaving them somewhat dry.

Following the suggestion found in an old cook-book for preserving cherries: "Put them with the sugar into close-covered crocks in an 'afternoon' oven (that is a gradually cooling oven), and keep them there for several hours," a fine result was reached as to flavor and tenderness, though the syrup required still further evaporation to reduce it to the jellying point.

FOR DRYING FRUIT.—If the tube in the top be raised to allow the free escape of steam the oven can be turned into a fruit dryer, the shelves being filled with plates of sliced fruit.

Various other uses as a warming oven will be suggested to any housekeeper. It is better than a *bain marie* for keeping food warm after it is cooked, and the top of the oven gives exactly the right temperature for raising bread, a tin pot lid being placed beneath to prevent direct contact.   MARY HINMAN ABEL.

Ann Arbor, Mich., September, 1891.

---

* The writer has two tinned copper vessels which fill the oven, one above another. In each of these, eight glass jars, of the type known as butter-jar, may be placed—making sixteen at one charge. This method of saving tomatoes without subjecting them to a heat above 200° F. is especially recommended.   E. A.

COMPARATIVE NUTRITION.

While this book upon "The Science of Nutrition" was going through the press, a beginning has been made in treating the subject of comparative nutrition.

When the twelve dietaries which have been given in the previous treatise had been prepared, it seemed probable that some use might be made of them in determining the relative cost of nutrition at the American standard in different states and countries. I could not of course expect to make anything but a crude beginning in this matter, because the habit of nutrition, if one may use this expression, varies greatly according to soil, climate, conditions and wages.

In countries where meat is scarce, the chief source of nitrogen is found in a large relative consumption of cheese or of beans or other legumes. How, for the price of a suitable day's ration may be equalized by the purchase of cheese or legumes in place of meat, remains to be dealt with. For the moment a few comparisons may be interesting.

It will be observed that in the twelve dietaries given, the Constants consisting of grain, vegetables and a modicum of butter or fat are uniform; they are computed in sufficient quantities to support life and are named

THE LIFE RATION.

The Variables, consisting mainly of meat, are given of different quantities, at different prices, and are named

THE WORK RATION.

All prices are given at retail for small quantities, except flour, which is assumed to be purchased by the sack or barrel.

Rations corresponding to Dietaries Nos. 1 to 4, have been computed in various places, with the following results:

Cost of 57 lbs. grain, vegetables and fat and 25 lbs. of the cheap cuts of meat, 82 pounds in all; sufficient for rations for 30 days, at 3467 Calories per day, the standard of a German soldier on a war footing being 3093 Calories.

Date. 1891.		Life Ration. 30 days.	Work Ration. 30 days.	Total. 30 days.
April.	Boston, Mass., U. S. A. (short crop of vegetables, 1890),	$2.31	$1.78	$4.09
November.	Boston, Mass., U. S. A. (vegetables abundant),	2.08	1.77	3.85
December.	Bismarck, N. Dakota, U. S. A.,	1.43	2.42	3.85
September.	Paris, France,	2.00	2.81	4.81
September.	London, England (in workman's section),	2.16	2.52	4.68
November.	London, England (in West End shops),	2.42	3.60	6.02
October.	Madison, Wisconsin, U. S. A.,	1.70	1.77	3.47
September.	Beyreuth and Nuremburg, Germany,	2.52	3.76	6.29
October.	Topeka, Kansas, U. S. A.,	1.54	1.53	3.07
December.	New Orleans, La., U. S. A.,	2.40	1.92	4.32
October.	Lincoln, Nebraska, U. S. A.,	1.38	1.78	3.16
October.	Ann Arbor, Michigan, U. S. A.,	1.87	1.86	3.73
1892.				
February.	Brussels, Belgium,	2.53	3.93	6.46
January.	Dresden, Germany,	3.14	3.30	6.44
March.	Munich, Bavaria,	3.30	3.63	6.93

	Life Ration, Cents per Day.	Work Ration, Cents per Day.	Total Cents per Day.
Boston, Mass.,	7.70	5.93	13.63
Boston, Mass.,	6.93	5.90	12.83
Bismarck, N. Dakota,	4.77	8.07	12.84
Paris, France,	6.66	9.37	16.03
London, England,	7.20	8.40	15.60
London, England,	8.06	12.00	20.06
Madison, Wis.,	5.66	5.90	11.56
Beyreuth and Nuremburg, Germany,	8.40	12.53	20.93
Topeka, Kansas,	5.13	5.10	10.23
New Orleans, La.,	8.00	6.40	14.40
Lincoln, Neb.,	4.60	5.93	10.53
Ann Arbor, Mich.,	6.23	6.20	12.43
Brussels, Belgium,	8.43	13.10	21.53
Dresden, Germany,	10.05	11.00	21.05
Munich, Bavaria,	11.00	12.10	23.10

The two returns from Dresden and Munich show the effect of the short crop of grain in 1891 upon prices.

I can of course claim only approximate accuracy for these comparisons. The personal equation will have a varying influence in obtaining prices. The habits of the people must be taken into view. In Boston, for instance, the tougher and coarser parts of beeves are sold for food; in Bismarck, North Dakota, they are probably put into the fat-rendering vats, not even being prepared for sale.

I have also been disappointed in the small number of returns received in reply to my circular, but yet hope to extend this inquiry, as there are now several associations in this country and in Europe which have taken up this matter in different ways.

Suffice it that even this beginning is very suggestive. It proves that where the nitrogenous element in food is abundant and cheap, labor is effective and wages are high. Where the nitrogenous element is scarce and dear, labor is not effective and wages are low.

Which is the antecedent and which is the consequent?

It will be observed that there is a much greater uniformity in the price of the Life Ration than of the Work Ration. May not this indicate a deficiency of nitrogen as the *cause* of low rates of wages?

Again, if one may venture upon a somewhat visionary hypothesis, another comparison may be made and another question may be asked.

Where the burden of Armies and Navies is heavy—nitrogenous food is scarce among the people—the army must be sustained even if the poor workmen starve. Witness Russia at the present time. Why must armies be sustained?

The army and the naval forces of the United States number only about 30,000 men and are all that we require. At the ratio to population of European armies and navies, making comparison only with the force in camp or barracks and paying no regard to reserves, our army and naval forces would number from 600,000 to 800,000 men; and since it takes the product of at least one man to support one worse than idle soldier, such a burden would be equivalent to setting apart ten per cent. or more of all the men of arms-bearing age from the productive and constructive work in which they are now occupied, to waste the most valuable and effective portion of their lives in the destructive work of preparation for war.

The mere money cost of this system of militarism which is eating away the vital force of most of the European States is about $1,000,000,000 a year. I am aware that army drill and discipline is justified as a mode of education. What does it cost? In Germany women do the scavenger work—sweep the streets—mix the mortar for the builders and perform the hardest work of the field, while the men, at the most productive period of efficiency, are obtaining the education thus claimed to be justified. France is a little better off, but many parts of Italy are worse.

It will be observed that the Science of Nutrition is not confined to the domestic kitchen or to the recipe book.

As the supply of food comes to half the cost of life or more, in many parts of this prosperous country, so the barbaric folly may be conceived of taxing the masses for the support of the classes by whom the military system is sustained and the military caste is supported in Europe.

In a recent prominent English review there appeared an article upon "The causes and effects of cruelty in conscript armies," which is significant, perhaps prophetic. When conscript soldiers who are kept under drill by cruelty are ordered to fire upon their own hungry brethren who are yet more oppressed, the guns may be turned upon the oppressors. It is claimed that European countries are prospering, but the national debts for war purposes are constantly increasing, and there is an increasing dependence upon other continents for the supply of food. These are bad signs, the classes may prosper more but how is it with the masses?

The foregoing statements of the relative cost of nutrition must indicate that the proportionate cost of food to other elements in the cost of living is less in the western states and cities than in the East or in Europe. In fact, the relation of food supply to the rate of earnings is even a more important factor than now appears. One can even predicate a rule on what is now known. It may be put in this form: To him that hath food in abundance shall be given the power to gain more; from him that hath not shall be taken even that which he hath.

These considerations bring into view the importance of the problem of securing a supply of nitrogen at less cost. Our Agricultural Chemists and Physiologists tell us that nitrogen is the most important, and at the same time the least abundant and most costly element in the nutrition of the plant, the beast and the man. The atmosphere is four-fifths nitrogen and we can't yet catch it: the iron smelting furnaces of this country are giving off ammonia enough to supply nitrogen to our fields, in large measure and we waste it. We have begun to save the phosphatic slag of the furnaces for use as a fertilizer, and thus to convert iron ore into corn and wheat— the next man will save the ammonia. In the meantime, Prof. Atwater and his associates are summoning the myriads of bacteria and microbes to our aid, who, living in their little dwelling places attached to the roots of clover, cow pea vines and other renovating plants, draw nitrogen from the air supplying the plant which, when turned under, renovates the soil.

In winding up this somewhat desultory series of treatises in which all the science has been contributed by my friends, I am reminded of my investigations in other lines and directions which have been previously published. I have been led by all my work to the conclusion that it is *Production which is Unlimited; Consumption which is Limited.* (See Introduction to the Industrial Progress of the Nation. G. P. Putnam's Sons.)

The hypotheses of Malthus in respect to population and of Ricardo and others in regard to diminishing returns from land, have as yet no apparent foundation in science or experience, and may be wholly discarded.

The present need in this country is not so much of instruction how to earn but how to spend an income, especially a small one. If the energy which corresponds to the present waste of food material could be spent for more adequate shelter, the evils of the slums would be abated and the bad tenements in the slums would be renovated. One may be warranted in estimating the present waste of food and fuel at five cents a day for each person, which is approximately twenty per cent. of the expenditure. To this we may add two cents a day, or less than half what is annually spent for liquors and tobacco—say seven cents a day in all. There are about 65,000,000 of us now, divided into 13,000,000 average families of five each.

At seven cents a day the measure of our waste of energy in converting good food into bad feeding, and upon liquor and tobacco at less than half their cost, amounts in each year to $1,660,750,000. If this waste of misdirected energy were converted into better methods of providing shelter, it would enable each family of five persons to spend 127\frac{75}{100}$ a year more for their dwelling places than they do now.

Suppose the waste only four cents a day, two cents on food and two on liquor and tobacco, even that comes to nearly $1,000,000,000—worth of misdirected energy—and the waste is greatest among those who can spare it least. The Imagination is the most potent force or factor in Religion, in Education, in Business and also in Economic Science and Statistics: its place in Literature is acknowledged.

How far I have succeeded in bringing Imagination, as well as a little Fancy into a Cookery Book, I must now leave to my readers to determine.

How much of this book may be attributed to a true Imagination, forecasting the near future when a good subsistence, comfortable shelter and adequate clothing will be so well assured to him or her who possesses health, aptitude and intelligence that it won't pay to be rich, and how much of it is mere fancy, the writer himself may not be able to tell. Suffice it that the following statements are of fact, and not of fancy or imagination.

Any intelligent boy can be taught in one lesson of one hour how to get up a five-course dinner for ten persons, consisting of Soup, Fish, Roast, Entrée, Vegetables and Pudding. I state this because I have done it.

Any intelligent girl can be taught to make and bake the best of bread by the use of the implements named in this treatise, without touching flour or dough with her hands. I state this because I have done it.

The fuel oil required for the cooking for a family of ten persons for one hundred days at the seaside in summer, can be bought at retail at less than one cent a day for each person. I state this because I have practiced it.

Any five intelligent women who can combine to live together in Boston in comfortable rooms in the new part of the city, can hire such rooms heated, with the service of a janitor added, at $62\frac{50}{100}$ per year each; they can purchase an adequate and comfortable supply of clothing at $62\frac{50}{100}$ each; they can purchase the food and fuel necessary to complete nutrition, with tea and coffee added, for $62\frac{50}{100}$ each, and within the limit of an income of $250 each one can apply $62\frac{50}{100}$ to sundries. I make this statement because I have personally verified every point.

I may not recommend this plan because those who possess the intelligence which would enable them to live in this way cannot perhaps afford to spend the time which would be necessary to such a plan, but will be able to earn more so as to spend more.

On the other hand, a great many of those who live dirty and ill-nourished lives in the slums could live comfortably and well upon their incomes, which in very many cases are ample, if they could only be taught how to spend them. My observations are limited mainly to the conditions of towns and cities. How far they can be applied in the rural districts and in strictly agricultural communities, I am unable to say.

In College Cities or towns in which the Corporation of the College supplies the land without rent, College Buildings can be constructed in suites of ten rooms, eight for dormitory and study, two for kitchen and dining-room,—as safe, as strong, as slow burning, as well lighted, as well heated and as well ventilated as the modern cotton factory is, which must possess all these attributes in order that the cotton may be spun, at a cost of not exceeding $800* per student—probably less. On this sum a rent of $62\frac{50}{100}$ per student would pay five per cent. interest, insurance, repairs and depreciation. I state this because I have made plans corresponding to the factories about which I know all the facts and have had estimates made upon them by competent mill constructors.

I have said that the education most needed now is how to spend, more than how to earn. I find as much evidence of this in the present expenditures upon Hospitals, College Buildings and Schoolhouses as I do in the conversion of good food into bad feeding. In the ordinary practice of combustible architecture, of which this class of buildings are apt to be typical examples, I find hospitals in which the inmates are exposed to cremation before they are dead, covered in with crazy roofs which do not keep out the weather; College Buildings which give the minimum of space and comfort at the maximum of cost, in which poor students can only be admitted by accepting charity; and finally Schoolhouses in which about two in three are bad types of separate invention on different plans, each more or less unsuitable to its purpose and costing from fifty to five hundred per cent. more than the sum for which a true typical schoolhouse can be constructed if the motive of the work be light, air and ventilation rather than outside appearances.

A typical schoolhouse can be planned and specified in interchangeable parts, so that orders could be given for four, six, eight and ten-room buildings of good exterior design and exactly adapted to their purpose within, with the same absolute assurance of minimum cost that has been secured in the construction of the textile factory, the paper mill and the machine shop of New England.

In the light basement of this schoolhouse nearly or wholly above ground, as a schoolhouse ought never to have a cellar under it, provision may be made for the service of the building, for manual instruction and for a Cooking Laboratory, in which instruction may be given in the simple principles of the Science of Nutrition and in the Art of Applying Heat under due control and regulation to the conversion of food material into nutritious food. This can be done without incumbering the premises with costly stoves or ranges, but by making very simple provision for cooking the food whatever it may be, in such a manner as to enable the boy or girl to carry back to the household such an example of the right method as to make the common practice distasteful. If the true kind of cooking apparatus, such as is used in the New England Kitchen, were set up, any kind of food could be dealt with, and in this way right methods might soon be brought into common practice.

With these dogmatic statements, which are neither the product of fancy or imagination, I will rest my case.

EDWARD ATKINSON.

BOSTON, March 1, 1892.

NOTICE.

A copy of this book will be sent to each person who has already bought an Aladdin Oven of the Standard size. It is possible that subsequent editions may be called for. In order to make them more complete and more attractive, each person who has discovered any faults, or who may have developed any merits in the oven which are not named herein, is requested to communicate with the undersigned.

New recipes or suggestions for the preparation of any kind of food are desired.

E. A.

TABLE SHOWING THE COMPUTATION OF DIETARY No. 1,

on the basis of which the comparisons previously given were made. It is one of a series of twelve dietaries of varying cost which are given in the test of "The Science of Nutrition."

CONSTANTS WHICH ENTER INTO THE SUBSEQUENT DIETARIES, NOS. 1 TO 12 INCLUSIVE.

ARTICLE.	POUNDS.	PROTEID.	FAT.	CARBO-HYDRATE.	CALORIES.	COST AT BOSTON PRICES 1891.
Flour,	22	2.64	.44	15.18	36,520	$0.55
Grain,	12	1.68	.84	7.60	19,800	.48
Butter,	2	.02	1.73		7,230	.56
Suet,	2		1.78		7,200	.12
Sugar,	2			1.93	3,600	.10
Potatoes,	10	.20		2.10	4,300	.25
Beets, Carrots, Onions, Squash, Cabbage, Parsnips,	7	.18	.03	.50	1,120	.25
For 30 days,	57	4.67	4.82	27.31	79,770	$2.31
For 1 day,	1.90	.155	.160	.910	2,659	.077

VARIABLES IN TABLE NO. 1, SHOWING METHOD OF ANALYSIS APPLIED TO EACH TABLE.

Beef, neck or shin,	12 (including waste)	2.00	.40		5,200	.72
Mutton, neck,	5	.62	.34		2,476	.30
Bacon,	4	.40	2.80		11,840	.48
Beef liver,	2	.40	.10		1,120	.12
Veal,	1	.19	.03		460	.08
Salt pork,	1	.03	.78		3,160	.08
For 30 days,	25	3.64	4.45		24,256	1.78
Total,	82	8.31	9.27	27.31	10,026	04.09
For 1 day,	2.73	.277	.309	.910	3,467.05	.136

The prices on which these computations are made were the retail prices in Boston, Mass., U. S. A., in the first six months of the year 1891.

TESTIMONIALS.

I may be permitted to present some of the replies which have been given to my questions. It will be apparent that I have taken all reasonable precautions so as not to mislead any one. It is my own opinion that this apparatus, or a better one corresponding to it, will ultimately work a complete revolution in the processes of cooking. Whether I am mistaken or not, will presently appear from the following letters which have been written by those who have made use of the ovens. They have been sent to me in response to the single request that those who have used the ovens would state to me their results and give the facts as they are. E. A.

In a report to the trustees of the Elizabeth Thompson Fund, Mrs. Ellen H. Richards gives her conclusions as to the requirements of an ideal cooking apparatus, as follows:

ESSENTIALS FOR GOOD COOKING.

APPARATUS.

For ordinary Cooking apparatus the following are essential points:

1. The degree of heat should be under perfect control; increased, diminished or withdrawn at will, and without loss of time. This can only be attained with liquid or gaseous fuel. Solid fuel demands constant and equable running and gives the best results when used in large masses. The small fire-box of a cook stove, and the urging of the fire for a short time three times a day are fatal objections to the use of anthracite.

2. A tightly closed vessel heated by steam, or hot water, or hot air, offers many advantages over the top of a red-hot stove or the inside of a nearly red-hot cast-iron oven for cooking, except for the broiling and the roasting of meat and for some other methods of cookery which require the quick application of heat.

3. For all purposes of slow cooking the oven should have a non-conducting covering which retains the heat where it is wanted, and also allows of tight closing and of security from the constant watching required by the fitful heat of a stove.

This use of a close oven with a non-evaporative atmosphere, seems to be the secret of the retention of the delicate and volatile flavors which usually flavor the house and street, and not the food as it is brought to the table.

The Aladdin Oven or Covered Stove. This is a square or oblong box of sheet iron of any desired size, with a non-conducting covering of magnesian cement or wood pulp, and is heated with a kerosene lamp or gas burner. The size in use for these experiments is 18 by 12 by 14 inches, and gives a cooking space at least equal to that of a No. 8 Crawford cook stove, and when empty can be heated to about 300° F. in an hour, and maintain that temperature for eight hours by a single kerosene lamp of the Rochester burner type, with the consumption of one quart of kerosene. When well filled with food materials in small portions the heat is sufficient to heat them in about twice the time allowed by an ordinary cook stove. When the space is completely filled with a vessel containing, for instance, forty pounds of meat and bone and fifteen quarts of water, the whole is raised from a temperature of 70° to 180° F. in seven hours, and to 212° in twelve hours. If the lamp is then taken away or allowed to go out, the temperature does not fall below 190° for four hours.

For this twelve hours, one and one-half quarts of kerosene are needed, or a gas burner can be used. For simplicity, effective use of heat, economy of fuel and development of flavor in the food cooked, combined with increase of its digestibility, the Aladdin Oven is an apparatus far exceeding in merit any other now in market. It will not meet all the demands that the modern cook now makes of the kitchen stove, and it may be in several respects improved, but in the application of well-known and long-tried scientific principles to the cookery of food, it is a distinct advance and a most valuable invention.

PORT ARTHUR, ONTARIO, CANADA, May 11, 1890.
EDWARD ATKINSON, ESQ., Brookline, Mass.

I gladly furnish my testimony in favor of your Aladdin Oven, for I feel that I can hardly say too much in praise of its numerous merits:

Its *economy of fuel*, not only in the limited amount actually consumed, and the low price thereof, but in the amount saved from waste whilst no cooking is going on, or during the slow simmering processes.

Its *economy of time and labor*, as the amount of requisite attention paid to the food during the process of cooking is so small; and there is no building of fire, and watching and regulating the same.

Its *economy of food itself*, for none of it need be wasted or spoiled, even by heedless people.

Its *cleanliness;* no ashes; no dust; no soot; no smoke from chimney or lamp.

Its *contribution to comfort*, not only as shown in above items, but also in preserving an even temperature in the kitchen or the house, and more especially during the summer, in escaping all extra heat; and also in the avoidance of steam, and the absolute freedom from the odors given off from the food in cooking, often very offensive, as from some vegetables, from corned meats, and the like; no smell from any scorched or burned food; no smell of kerosene smoke, nor of coal gas.

Its *convenience* in the *instantaneous* starting and extinguishing of the heat, and the increasing or modifying of the same, without the trouble of attending to the fires, etc., etc.

Its *convenience* also in having everything handy at meal times, as the oven, whilst in actual use, can be kept within arm's length of the dinner table, without the least offense of any sort to the family or guests.

Its *convenience* also in lighting up the apartment, if desired, at the same moment that it is doing its proper work of cookery.

Its most notable *convenience* to early risers, laborers and others, or to nursers or watchers, in doing its work *automatically* through the night, and furnishing a hot breakfast at any desirable hour, all ready to hand; or in the same way serving bachelors, who must be absent during the day, to find a hot meal awaiting their return.

Its *ministering* to the *pleasures* of the *gastronomer* or to the delicate palate of the *invalid*, by giving such marked improvement to the flavors and the *tenderness* of fish, flesh and fowl—of breads, cakes and other viands. Many old dishes acquire a new charm in this oven.

Its *conducing to the health* of those who use it (and this *should* be the highest consideration of all) through the superior quality of the cookery, which is undeniably rendered more wholesome and nutritious under this slower, gradual and thorough process. It must tend largely to obviate or to mitigate the horrors of *dyspepsia*.

Its *compactness* and *portability*, making it available for travel, and just the thing to carry with one to summer resorts, or for permament use in camps, and frontier cabins (where my own is now, and of invaluable service).

Such a combination of advantages over any cook stoves or ranges, over kerosene stoves, or even gas stoves, ought to give the *Aladdin Oven* a preference over all, and bring it into general use among all classes; but even more universally among the poor than the rich.

The inventor disclaims the ability or *desire* to *fry anything* in this oven. For my part, I confess to being fond of a fry. And by some management one can get up a pretty good *substitute* for it, in the oven.

In its present form there is no special arrangement of utensils for hot water. But some water can readily be heated on the top of the oven, whilst cooking is going on inside, by setting a vessel, properly fitted, over the vent-pipe.

Very truly yours,
(Signed) EDWARD A. WILD, M. D.
Late Brig. Gen. U. S. Vols.

THE STORRS SCHOOL AGRICULTURAL EXPERIMENT STATION.
W. O. ATWATER, Director.   C. S. PHELPS, Vice-Director.   C. D. WOODS, Chemist.
MIDDLETOWN, CT., June, 1890.

To whom it may concern :

The Aladdin Oven has been in constant use in my family for the past five months, and in that time, with all the cooking that has been done in it, there has been no such thing as complete failure with any of the dishes prepared and cooked in it, although these were largely experimental in their proportions and especially in time of cooking. At first some miscalculation as to the amount of heat given by different lamps gave rise to some uncertainty in time required for cooking.

Bread baked in this oven acquires a peculiarly agreeable nutty flavor, such as I have never found in bread otherwise baked. Cereals of all kinds, cooked for a long time at a low temperature, are much superior in taste to those cooked in the ordinary way. Cheaper, and consequently tougher cuts of meat can, by the application of low heat for a long time, be cooked so as to become as tender and juicy as the better cuts are as ordinarily cooked. Roasts of meat, especially lamb, veal, pork, poultry, etc., have their flavor greatly improved by this method of cooking.

(Signed) CHARLES D. WOODS.

JAMES C. BOYCE,
Attorney-at-Law and Solicitor of Patents, and Counselor in Patent Causes, 91 Water street, Pittsburgh, Pa. (With Oil Well Supply Co., Limited.)

PITTSBURGH, PA., May 5, 1890.
EDWARD ATKINSON, ESQ., Boston, Mass.

Dear Sir :—Replying to your letter of April 22, I would say that we have been using your Aladdin Oven for several months. Too much cannot be said in its praise. It does its work to perfection, and everything cooked in it is tenderer and more finely flavored than anything I ever tasted before. We use it with natural gas. The consumption of gas is very trivial. At the rates charged for natural gas, the closest calculation that I can make shows that it does not cost over five cents per month for cooking for a family of five persons. It uses about three feet an hour, or about ten feet a day, say three hundred feet a month.

Much as we have liked it heretofore, it will be still more valuable during the summer months, as the heat is concentrated around the oven, and is not radiated through the room. It will, therefore, avoid all the discomforts of cooking in the summer time.

I shall be happy at any time to have you refer to me, and will cheerfully show the operation of the oven to any one who will visit my house at the corner of Fifth avenue and Halket street.

Yours truly,
(Signed)   JAMES C. BOYCE.

NEW ENGLAND KITCHEN, 142 Pleasant street,
BOSTON, July 15, 1890.

MR. EDWARD ATKINSON.

Dear Sir:—For the past six months we have had in constant use in the New England Kitchen, three Aladdin Ovens of the largest size. In them we have made daily from fifty to seventy-five quarts of beef broth and twenty quarts of pea soup. We did not settle on your oven for this purpose until we had thoroughly tested it and all other utensils that seemed likely to meet our requirements, which were that the broth should be invariable in quality from day to day, and that as large a quantity as possible of a fine flavor should be yielded by a given weight of meat, and that also with a small outlay for labor and fuel. Your oven meets these requirements exactly. The cost of fuel for making twenty-five quarts is five cents, and no care is needed between preparing the meat and straining the broth when done.

We recommend this method most heartily to hospitals, hotels and all institutions that cook in large quantities.

MARY HINMAN ABEL.

HERBERT HALL, WORCESTER, MASS., Sept. 25, 1889.

To whom it may concern:

This is to certify that I have had the Atkinson Oven or Baker in use at my house and have quite fully tested its merits.

Vegetables and meats were cooked in the oven in my dining-room without giving off any perceptible odor.

The cooking was exceedingly well done, without scorching or drying. The meats were tender, juicy, and of unusually good flavor.

Puddings and pies were not only well cooked, but were ripened, so to speak, as, it seems to me, they could be by no other process with which I am acquainted.

The work is accomplished without care or concern and at a merely nominal expense.

Food is not only much more cheaply cooked by this process, its value is greatly enhanced. It is more nutritious, and it is more digestible, thus promoting health and longevity.

It is not easy to estimate the full benefit which might be derived from the adoption of this cheap, wise and healthy mode of preparing food.

(Signed) MERRICK BEMIS, M. D.

---

23 ARLINGTON STREET, WEST MEDFORD, MASS., July 23, 1890.
EDWARD ATKINSON, ESQ.

Dear Sir :—My knowledge of the merits of the Aladdin Oven leaves little doubt in my mind that you must have already received such favorable opinions of it from high places as to put my lame excuses into the shade. It has occurred to me, however, that my evidence as a working man might have a peculiar value on that account. My oven measures ten inches each way inside. With the assistance of a little single-wick oil stove for boiling, it does the entire cooking for my family of three. Our breakfast, consisting usually of oatmeal porridge, stewed fruit, and perhaps fish, or some simple meat stew, is invariably put into the oven before we retire for the night, and is always perfectly cooked in the morning. The coffee on the top shelf has a fine flavor when treated in this way. The oatmeal, etc., is put on the lower shelf and the lamp turned down until the flame is about half size. Early on the morning of July 4, we left home on a visit, having first placed the usual breakfast in the oven, and put under it a small lamp turned low enough to make the oil last. We arrived home at eight o'clock the next morning to find everything cooked as nicely as usual. And what was more important just then (we being hungry after traveling) the food was at just the right temperature to commence operations immediately. I could add, if it were necessary, many other occasions when we made extraordinary demands upon the oven, never to be disappointed.

It is, in my opinion, the only cooking arrangement which has yet been invented which does its work with absolute certainty and exactness.

Very truly yours,

WM. P. PERRY.

Dear Mr. Atkinson:
I am thoroughly convinced of the justice of all your claims for the Aladdin Oven, and you may use my name with the greatest pleasure on my part. I am sorry that I cannot be set down as one who has discarded all other means of cooking save this; but the fact is that the oven remains only an adjunct to the range in my household. This is only because I have not yet done sufficient battle with my cook, who is conservative by temperament, illogical by sex, and unable to see her true interests by race. But I felt last summer that I was tasting, for the first time, certain vegetables long familiar by name.

Very truly yours,
ELIZABETH FAIRCHILD.

LOWELL, Jan. 8, 1891.
MR. E. ATKINSON.
Dear Sir:—After a year's experience in the use of the Aladdin Oven, I am most happy to bear testimony to its value as enhancing the nutritive qualities of the food cooked and to its reliability. Once having determined the time required for cooking an article, the same result is obtained every time without watching or anxiety. For preparing soup-stock, cooking corned beef, tongue, or the various sauces, apple, etc., it is perfect. I am frequently asked if fancy articles can be cooked in it; to which I would reply that for angel cake, popovers, sponge cake, custard souffli, I would use no other unless obliged to. You are at liberty to use my name if you desire. I consider the oven invaluable in my family, not only from an economical point of view, but as a labor-saving invention.

Very truly yours,
(Mrs.) M. A. SARGENT.

66 MARLBORO ST., BOSTON, Jan. 12, 1891.
MR. EDWARD ATKINSON.
My Dear Sir:—I had your Aladdin Oven in use in a farmhouse in the White Mountains during last summer, as an annex to the customary wood cooking stove.

I found the Aladdin Oven an ideal one for—First, economy in fuel; second, for dispensing with the intense heat of a cooking stove; third, for the saving of labor in watching the process of cooking,—thus, beans and oatmeal may be cooked all right with no attention from the cook; fourth, and to me most striking and important of all, for producing by its slow process of cooking a better flavor. This remark is applicable to soups, fish, corn and other meals, rice and bread puddings and fruits.

I shall be glad another summer to make further experiments; not only a pliant and receptive disposition, but time is needed to evolve the full capabilities of the Aladdin Oven.

You are at liberty to use my name as one interested in any experiment which may contribute to the simplicity, economy and nutritive value of the home-table to the family.

I am sincerely,

EMILY TALBOT.

The Aladdin Oven economizes the housekeeper's time wonderfully, for by its aid many processes of cookery may be left to care for themselves, which with a range would need constant attention. My experience with it was mainly in the summer time, and I found it possible to put fish or meat, puddings and vegetables side by side, and go outdoors, knowing that dinner would be ready when wanted.

I have also used it with great success for extracting and clarifying fat, for baked beans and brown bread, cooking fruits and marmalade, etc. From my experience with the latter I see no reason why all the family canning should not be done with this oven. The fruit is thoroughly cooked *without burning*, and retains its shape.

A housekeeper who tries to apply her experience with a cook stove to this oven might at first be unsuccessful. New standards must be adopted, but this once learned, with one of these ovens and a small oil stove for quick cooking, no one would wish to change again for a range with its dust and ashes.

The next generation of housekeepers, who to-day are being trained in the cooking schools, will have discovered quick cooking does not develop digestible and nutritious food. ANNA BARROWS,
Teacher of Cooking,
North Bennet St. Industrial School.

---

106 HAMMOND ST., CHICAGO, Jan. 12, 1891.

Dear Mr. Atkinson :—It affords me great pleasure to answer your letter of the 5th, because I am able to bear testimony to the value of the Aladdin Oven, and especially because Mrs. Trumbull desires me to add her testimony to mine, and indeed hers is the most important.

She thinks that you will be classed among the great abolitionists in the domain of domestic economy; that, as the abolitionist of cook-stove slavery, you will be famous as Elias Howe, the abolitionist of needle slavery ; and in due time, she hopes that there will arise in Boston or thereabouts an abolitionist of wash-tub slavery to complete the triumvirate.

It is only fair to say that Mrs. Trumbull was for a long time sceptical as to the value of the oven ; for the reason, I suppose, that it was so easy, simple and cheap. For several weeks it was neglected under the pretext that we had no proper lamp for it, and that some day, when down in the city, she would buy the exact lamp required, and give the oven a fair trial. Your address to the doctors excited her curiosity, and she determined to test the oven by the use of an old "Daylight" lamp, which we found among the rubbish in the cellar. The result was a great triumph for Aladdin. The cooking was perfect, and the fuel cost scarcely anything, while the ease and comfort of the cook gave a calm, benign, religious tone to the temper, unattainable under the cook-stove reign.

Reckless of the fate of Thomas Gradgrind, you exclaim, "Give me facts." Very well, I will give you the facts of yesterday's dinner. We had a wild duck and roasted potatoes, escalloped potatoes, rice, maccaroni, and grape pie. These were all put into the oven together, and we had a shelf to spare. Then, as we have no hired girl, this important question arose, "Dare we all go to church and leave our dinner to the sole care of that old Daylight lamp ?" It was decided to risk it, and the whole family went to meeting, leaving the dinner and the lamp to fight it out alone. On our return we found that the lamp had remained faithful at its post, and every article in the oven was perfectly cooked ; no essential property of the food nor any flavor of it lost. Afterwards, and as a matter of mere pastime, the oven baked for us "four lovely loaves of bread." The adjective is not mine ; I tell the tale as the tale was told to me ; I quote the very language of the report as it was made to me by the autocrat of the kitchen table, "four lovely loaves of bread."

I have reserved our greatest triumph, "Beans," for my final testimony. It has been our custom in the winter time to have beans for Sunday breakfast, cooked a la Bostonnaise ; and to cook them thus and do it well is high art, a feat requiring native talent to perform it, as you very well know. My wife had acquired the accomplishment, partly by hereditary transmission, and partly by special training in New England. Last Saturday she took the desperate risk of trusting our Sunday jar of beans to the skill and integrity of Aladdin. The lamp was on duty all night while we were all asleep, and in the morning the beans were done to a degree of softness, delicacy and flavor never attained before ; and what is of great importance, "without the waste of a bean either at the top or bottom of the jar." I quote again literally from the report of the autocrat.

It is our intention shortly to give a dinner to a few sceptical friends, every article of the "Menu" to be cooked in the Aladdin Oven.

You are at perfect liberty to use my name and testimony in any way you please, and my wife says, "Tell Mr. Atkinson he may use my name also;" therefore, I will sign this letter with both our names.

Most truly yours,
M. M. TRUMBULL and FRANCES TRUMBULL.

To HON. EDWARD ATKINSON, Boston, Mass.

NEW YORK, Feb. 24, 1891.

EDWARD ATKINSON, ESQ., Boston, Mass.

Dear Sir:—Your favor of the 9th inst. came duly to hand, but pressure of duties has prevented my attending to it before this. Please accept my apologies for the delay, but it has been unavoidable.

In regard to the Aladdin Oven, we have used it for a number of months, and although we had been led to expect much of it before we received it, I am frank to say that it has more than met our expectations. The more we use it the better we like it, and we have given it a very thorough trial. Except for the few articles which require a sharp, quick heat, we find it more satisfactory in results than any ordinary oven in our knowledge. We have cooked meats of all kinds, vegetables, puddings, pie, etc., in fact all foods save those requiring quick heat, and in every instance the oven had proved itself an unqualified success. We think that food cooked in the Aladdin, meats especially, has a better flavor and pleasanter taste than the same cooked in the ordinary oven. My wife and I are very fond of Boston baked beans; we have cooked them in the Aladdin and we believe they have a richer and better flavor than ever before. We have placed different kinds of food, chicken, potatoes, turnip, onions, or pudding in the oven, started the lamp and then left the oven to "cook itself," leaving the house entirely for four hours, and coming back found everything "done to a turn." We find that the lamp can be so regulated as to cook a dinner in two hours, or four or five as you choose; the food is just as nice whether it be done in two or five hours. This is a wonderful help when one does not keep a servant; the housewife is not tied to the kitchen stove, but can go visiting or receive visitors without the worry of a possibly ruined dinner, or start her dinner and then turn her whole attention to something else.

The thing not the least worthy of commendation is the arrangement of shelves in the oven. We find that articles which in the ordinary oven used to be cooked the longest, should be placed on the bottom of the Aladdin Oven, graduations between. The heat of the Aladdin is so distributed that this ensures even and satisfactory cooking of all. The expense of running the Aladdin we estimate at not more than one and one-half (1½) cents per hour, rather less than more.

After several months' experience with the Aladdin Oven, my candid judgment is that it is almost invaluable to those living, for instance, as many do, in cities, I mean under what is termed "light housekeeping," although it is good in many or all homes. The Aladdin is clean, compact, always ready, requires no attention when once the food is put in, and it serves as a most excellent heater for a room. During the winter we have used no other stove in the room where the oven is, and needed none.

You are at liberty to use any or all of this, as you choose.

Yours very truly,
L. W. RAYMENTON.

TOPEKA, KAN., Feb. 4, 1891.

MR. EDWARD ATKINSON, Boston, Mass.

Dear Sir:—Your favor of the 16th ult. duly at hand. We can say that we are well pleased with your oven and consider it the best thing we know of for our use. We keep it in our sampling-room attached to the office, where we can bake a batch of bread at any time. We made one change. viz., getting a larger lamp so that we bake the bread in the time of an ordinary stove. The writer has experimented with making bread as you suggest, raising three hours, and baking three hours with the small lamp, taking it home where we like it better than the other way of making bread; although it is not so fine-grained, perhaps the fault of the baker. Have also tried a beef roast with the small lamp, and a better cooked or jucier and tenderer roast never came on our table. We intend having one of the ovens in our kitchen before hot weather comes.

Yours truly,
CROSBY MILLING Co:
T. R. E. KIDDER.

You can use this for publication if you wish, as we state only facts.

ST. LOUIS, MO., 2719 Chestnut St.

MR. EDWARD ATKINSON.

Dear Sir:—Your very kind and satisfactory letter of April 28th was duly received. I want to tell you that our oven is every day becoming more indespensable in my house. The servants say it reduces the work one-half, and I feel sure they would find it hard to go back to the old way of keeping the range burning all day. The kitchen is cool, clean and quiet. We make use of the gas stove for sauces, tea and coffee, and of the charcoal-broiler attached to the range, for broiling and toasting, although the gas stove is capable of these operations, if liked, but I am epicure enough to believe that *perfection* in this direction can never be reached, except by means of hot coals, and it is very little work, if a servant will use her brains at all, to light that little pan of charcoal. It only means to think *in time*, and this is the secret of success with the Aladdin Oven. The only difficulty I experienced is the space in the oven is not enough for all I want to do. But I shall send a check for a second oven probably next week. Then I shall be quite complete, and shall be glad to know when any larger sizes are made.

I send you a few items of my own experience. I have a large pail made of heavy tin with cover, holding four gallons, which I keep filled with water on the top of the oven (taking out ventilator, etc., of course), and the water keeps hot enough nearly all the time for washing dishes not very greasy, etc., so none of the heat from oven is not utilized. Soup for the family is always made at night. One even teaspoonful of salt to five pints of water, six whole black pepper corns and one bay leaf, put in at night when the water is cold, soup bone and meat, onions and vegetables. In the West we are not afraid of onions; we use the flavor wherever it is needed.

It requires fully two weeks of constant use to get the oven seasoned, then this list can be relied upon, we think, as to time:

Roast beef (over 8 lbs.),	25 minutes to pound.
" " (under 8 lbs.),	20 " " "
Hindquarter Spring Lamb,	2¼ hours.
Smothered Chicken,	8 "
Roast Chicken,	3½ "
Chicken Pot-pie,	3 "
Irish Stew (breast and shoulder),	3 "
Cauliflower (put in boiling water),	1½ "
Asparagus (" " " " "),	1½ "
Sliced raw Potato Stewed in Milk,	3½ "
Seasoned when put in with salt, pepper and butter very nice.	
Baked Apples,	2 "
Stewed Prunes,	2 "
Graham Bread (8 in. x 3½ in.),	4½ "
White Bread (8 in. x 3½ in.),	3½ "
Layer Cake,	1 "
Corn Meal Muffins,	1½ "
Best made with buttermilk and soda, rather than made with baking powder.	

I ought to have a larger list, as we have cooked all kinds of vegetables, but the fact is, that the cook is now getting so used to it, that she begins to use her judgment instead of the list, and so unconsciously forgets to put it down, as she did at first. I am also using patent cake pans, that require no greasing—a great saving of time and labor and butter. I hope I have not written you too long a letter. I shall be in Boston in July.

Yours very truly,
(Mrs. E. C.) CORDELIA S. STERLING.

From the same correspondent I have received, at a later date, the following additional testimony:

St. Louis, June 17th, 1891.

Mr. Edward Atkinson.

Dear Sir:—To be able to carry on the necessary operations for the family in cool rooms, without the oppressive heat which is customary in kitchens and laundries, is a revolution in domestic economy. I think it is almost a drawback that one can claim so much for the oven. It makes people incredulous, as I was at first. I use round pans or dishes, the blue and white enameled German ware, which is pretty and neat. If one could get them in the shape of a parallelogram, like a pan for baking bread, three could be put on one shelf, with a roast of meat on the lower shelf; the next above must be omitted. With a pan of vegetables five inches deep, the top one must be left out; so we find in practical use only one shelf and the bottom of the oven available. You will have to provide for the manufacture of cooking vessels exactly adapted to preserving every inch of space. This I know is not as interesting to you as the question of saving every atom of nutriment in the food cooked.

I have been making some experiments with breads and cake made with yeast powder, which is supposed to require a very hot or rather quick oven, and with perfect success. Strawberry shortcake was delicious; also I have made some experiments with fruit. My children like gooseberry jam for school lunches. I put ten pounds of gooseberries and six pounds of sugar into a yellow-ware jar (butter jar) with close cover, a half teacupful of water on the bottom of the jar, then the fruit and sugar in about three alternate layers and put on the cover. Put the usual iron slide on the bottom of the oven and a slide with holes upon that; the height of the jar admitted of no shelves. This was put in the oven at 9 p. m., when the cook went to bed, and the lamp lighted. When she came down at 6 a. m. she extinguished the lamp, but did not open the oven (according to my order). At 8 o'clock I opened the oven and took out the jar. It was perfect; tipping the jar from side to side the contents were almost a jelly, the fruit scarcely broken and a beautiful claret color. I put it immediately in the glasses and the work was done. Contrast this with the five or six hours watching and stirring over a hot stove, occupying the greater part of a day with the operation. To-day I have canned three gallons of strawberries and made one dozen glasses currant jelly, and have not been in the kitchen over, if as much as two hours all together. No assistance from the cook except in preparing the fruit. The strawberries were placed in the above mentioned jar in alternate layers with the sugar (no making of syrup beforehand), and left in the oven four hours. The jars were covered; once during the time the cover was lifted and the fruit pressed down under the liquid with a wooden spoon. At the end of four hours the jars (the ordinary glass fruit jars) were placed a few moments in the oven to heat them, and immediately filled with the mixture. The fruit is perfectly unbroken, and is perfect in flavor. As to the currants, after the currants were picked from the stems, the fruit was washed, rubbed and thoroughly macerated to free the seeds from the coating of gelatine which surrounds them, which is indispensable to the making of jelly. This is the laborious part of the process, and constitutes all the "science" of making currant jelly. The mashed fruit was then put in the oven, and reached boiling point in one hour, but was left one-half hour longer, then strained and measured, and pound for pint of sugar added. At the end of an hour and a half it was taken out, found to be done and put in glasses. I have had a little cooled on ice, and I feel sure it is successful, and all this time the kitchen is as cool and as free from confusion and sticky dishes as possible, as if nothing special was being done. Pardon the long letter. I am much interested in the pail, etc.

Yours truly,
CORDELIA S. STERLING.

---

* Having been attracted by an extremely intelligent report of the cost of subsistence in a workman's family, given in Commissioner Carroll D. Wright's report upon the "Iron Industries," I obtained the correspondent's address from the Commissioner, and happening to have an oven lying in Chicago, which had missed its purpose, transferred it to her; being well assured that the report of such a truly accomplished woman would be of great value, especially as she might give a sound opinion of the value of the apparatus in the family of a workingman earning the ordinary rates of wages in the iron industry.

Her report gives the *outs* as well as the *ins*, and also represents the slight difficulties which may occur in the first effort to apply the oven to use, especially as she had overlooked my careful statement that "potatoes and other tubers require a higher heat than any other articles of food, and must be specially dealt with if success is expected."

CHICAGO, March 25, 1892.

MR. EDWARD ATKINSON.

Dear Sir:—I am glad to say that I think the Aladdin Oven is a treasure to the woman who likes slow baking, good roasting, and—*no worry*. I think I have not given it the test it deserves, because I have had the range "going," and have not formed the habit of beginning *soon* enough to get a meal. We of the cook-stove service find it hard to get used to the *long time* it takes to cook a thing in the oven— and, I have not felt that I could afford to *experiment* on anything but the foods we habitually feed upon. So far as I have used the oven, I think it is the most valuable kitchen accessory I ever handled; but I have concluded that it is unreasonable to expect it to do *every*thing as I thought it would, when I first read your pamphlet.

My experience is, that it will not do *any*thing that should be done *in a hurry*. Whether the "hurry" may be dispensed with altogether, is still to me a problem, though I am disposed to favor the negative.

I am a devotee of the frying pan. I think the poor thing has been very much abused, and I think you must have been much abused by it, or you would not speak of it in the terms you do. It is, indeed, an instrument of torture in the hands of the careless. The cook stove has displaced the fireplace, to the detriment of certain modes of cooking, notably broiling and toasting.

The oven may displace the range, but it seems to me that we shall miss the *fry*, the chief charm of which is the toothsome brown that comes of *quick* closing of the pores, thereby keeping in the juice. If that is unwholesome, I have not yet found it out.

I am fond of beautifully-browned baking-powder biscuit. I have tried every arrangement of time and condition I could think of, with anything but desirable results. Mrs. Trumbull's efforts met the same fate. If it can be done in the oven, I shall be very glad to learn how.

I don't think I have anything to say in disfavor of the oven. I believe in *specialties* rather than *cure-alls*, and I take the oven as a specialty.

I was struck with your pertinent recipe for cooking, i. e.—" One part gumption," etc.—and wondered if it would be *more* pertinent—when the oven was "seasoned." To work two or three weeks, not knowing whether a series of *terribly flat failures* is caused by a lack of gumption, or an unseasoned oven, is—well, not *very* inspiring.

I have thought perhaps you might subject the ovens to a process that would fit them for *immediate, successful* cooking. That would make them a *benefit* from the start, instead of an *experiment.*

I have tried the "all-night cooking" but once, and that, with a flat wick lamp, was *such* a failure I have since waited till morning to get breakfast, though I intend sometime to try the "Rochester," burning low.

My first effort was baked potatoes; they were small and would have baked in thirty minutes in the cook stove, but at the expiration of one hour and each of the three succeeding hours when I tried them, they were *as hard as rocks.* Since then I have had large potatoes perfectly baked in three hours. Eggs, broken into a hot dish, and seasoned, were good in ten minutes. Tomatoes, one and one-half hours—flavor superior. Pork and beans, six hours—delicious. Breakfast bacon, one hour—*very* nice. (Turn the meat once within the time.) A five-pound "chuck" roast, three-quarter hours to the pound, and sweet potatoes—perfect. Green coffee, three hours—(not much in the pan to make sure of my first effort). Any one who browns her own coffee knows what a triumph that was—no smoke, no stirring, no anxiety lest it burn. Baked apples, three hours—fair. The quality of the fruit to be considered. Light bread, three hours—good, but not nice looking. A shoulder-cut of beefsteak cooked nicely in ten minutes, turned once. A cake ten inches across the top and about three deep, baked in two hours. I have not yet tried to boil anything.

Thus you have my limited experience with the Aladdin Oven. If it is of benefit to more than myself, I shall be glad. I should like to give you the comparative cost of food and fuel with, and without the oven, but cannot till I put it into constant use. It would take several months to obtain a fair statement, but I think it would be interesting. Clothing and shelter is another problem in economic living, which has, as yet, remained beyond my powers.

Do you recommend kerosene for dishwashing, from your own knowledge? I was told once to use it for the family wash—that it would "take the dirt right out"—but I found, as others have, that it took one-third *more* soap *to take out the oil;* and concluded not to give the credit for work to kerosene that soap would do by itself. Thereafter I used one-third more soap in my washing and left the kerosene out. A little *more* "lukewarm" water will rinse off the suds made by an extra amount of soap in the dishwater, but it will not take away the taste and odor of kerosene—at least, not for me.

CHICAGO, April 4, 1892.

MR. ATKINSON.

Dear Sir:—I wish to *add* to my letters of last month concerning the oven. I have, since then, had *success* with light bread; it was nicely browned, and sweet.

And I had a "boiled" dinner that was the best I ever ate. The surprise of it was that we had a perfectly-cooked dinner in a little more than two hours, when the same food would have required nearly four hours on the cook stove. Hereafter my boiled dinners shall have a *steady* heat under them, and not be uncovered till they are supposed to be done. My admiration for the oven *grows* with each day's use. Who knows! Perhaps I may learn to "fry" in the oven.

Kindly yours,

COLUMBUS, GEORGIA, June 19, 1891.
MR. EDWARD ATKINSON.

Dear Sir :—The Aladdin Oven has been in daily use in my family for about two months, and all the various kinds of meats and vegetables in the market have been cooked. The results have been so highly satisfactory that every member of my family joins in insisting on everything being cooked in it. The improved flavor given to food by this slow process is very manifest.

For meats of all kinds, it is unequaled. That verdict has been unanimous by every one of the numerous persons who have tested it; meats, which from an ordinary oven would be tough and undesirable, from the Aladdin are always tender. Southern mutton, usually as tough as dried beef, becomes tender as veal. In one instance, to test this feature, my wife bought a cut of neck beef, which the butcher said could not be eaten, and protested against the purchase; but after cooking, it was tender as chicken, and would have been pronounced first-class roast beef elsewhere. I believe that any kind of meat, not allied to gristle or leather to start with, will come out of this oven tender. Have never yet found any portion of any meat cooked, not thoroughly tender and of good flavor. No one will believe what a change it makes until convinced by actual trial. Another feature is, that while in an ordinary oven there is considerable loss of weight in meat in cooking, in this one there is very little loss, and almost as much weight comes out as went in. With the various kinds of vegetables cooked, my family agree that all, except Irish potatoes, are better from the Aladdin, and some are very much better. A great variety have been repeatedly tried, as every day the oven has been filled with meat and vegetables for nine persons. Cooked in closed vessels the loss in weight or natural flavor is small, and vegetables come from the oven with natural taste remaining and in the most desirable condition.

Cost of oil for heat is very small and much less than fuel for a stove. We like the oven so well, I am building a special small room especially for it, and would not dispense with its use under any consideration. Any one who will try will have a like experience. No previous instruction in use is necessary—simply prepare the food, put it in the oven over the lighted lamp, and the oven itself will do the rest—requiring almost no attention. Time will demonstrate the very great value of this system of cooking. It has passed the theoretical stage. Trial demonstrates to any one the results.

Yours truly,
JOHN HILL.

FROM A WESTERN LADY.

DEAR MR. ATKINSON:

I doubt not you are frequently met with the objection, that servants are not willing to use the Aladdin Oven, and my own experience has been that any attempt to reduce labor or waste is met by an amount of prejudice and obstinacy which is extremely annoying, and which is submitted to by intelligent housekeepers who would not show a similar subserviency under any other circumstances; but a moderate amount of superintendence and explanation combined with the successful results which never fail, will soon convince the most obdurate. A Canadian French woman, who had followed the business of cooking for more than thirty years and who was thoroughly steeped in the platitude that "old ways were good enough for her," affords an illustration of what can be done by the argument of facts. More than the ordinary amount of insistance was necessary on my part to compel the use of the oven. But she yielded sooner then give up her place, which was made the alternative. Before the end of a week, she would not willingly have gone back to the range. Before the month was out, a friend having lost her cook and having a case of sickness in her family, I desired Marie to go to her, promising her a good place and where she would meet with every consideration she ought to expect—especially as my family was leaving for the summer home and I could not keep her much longer and she could remain with Mrs. S. indefinitely. The transfer was made and all went well for forty-eight hours, when my friend was surprised to learn that Marie wished to leave—her only explanation being—"I have decided not to cook any longer over a hot range." Expostulation was in vain—her only reply was—"People must get them kind of ovens if they want to keep cooks—These ranges they can't stand in summer"—and so she left.

With regard to the incident of the cook, I am only too glad to furnish any proof that no obstacle need exist among servants, against using the oven, if mistresses are willing to once set them on the right track—to look after the business of their households, as a man looks after the business of his office.

BROOKLINE, June 16, 1891.

EDWARD ATKINSON, ESQ.

Dear Sir :—I am very glad to be able to testify in favor of the Aladdin Oven. The cook likes it, as it simplifies work in the kitchen and the kitchen itself is kept comparatively cool in this hot weather. The family likes it on account of the great excellence of the food. We never think of having our meat or bread cooked in any other way.

Both of my children formerly ate meat, beef and mutton with reluctance, had to be urged, and only a little could be given them. Since using the oven they have taken a great fancy to meat and seem disposed to make the entire meal of it. This alone is sufficient justification for the oven.

So far as we have tried it I am able to endorse all your own encomiums. In fact, we have frequently put an old fowl into the oven and turned him into a spring chicken.

You are very welcome to make use of my name if you wish.

Very truly yours,

TUCKER DALAND.

[From the Monthly Bulletin of the State Board of Health of Iowa.—J. F. Kennedy, M. D., Editor.]

THE ALADDIN COOK STOVE.

There are a few things that everybody does—that they *must* do to live. One is sleeping, another eating. We wish briefly to speak of the latter. While many apparently live to eat ; all necessarily eat to live. The subject of all eating is, or should be not only to satisfy the demands of hunger, but to afford the means by which the digestive organs, supplemented by the circulating blood, may restore the waste that is continually taking place in our bodies.

In order that the stomach and other digestive organs may most promptly and efficiently perform their important functions, the food taken should be in a form as palatable and easily assimilable as possible ; as well as nutritious in quality. To be thus palatable and assimilable it is highly important that the food should not only be as perfect as possible for its kind, but be prepared properly. The most important factor in the proper preparation of food is its cooking. There are many devices for cooking—some very primitive and crude—some modern and scientific. Proper cooker as an art and a science, and to properly practice it, requires not only training but suitable appliances.

Without going into details we believe the apparatus that best preserves the juices and aroma of the food, and at the same time renders it tender and easily masticated and digested is the best—especially if at the same time it does it most economically.

We are not personally acquainted with all the various cooking appliances in use, but we have never seen anything that so nearly comes up to our idea of perfection in a cooker as the Aladdin Cook Stove.

The inventor is the Honorable Edward Atkinson of Boston—a large manufacturer there; and the invention is the result of much study and many years of experimentation with the object of finding for his employees something that would most perfectly from a sanitary standpoint, as well as most economially serve as a cooker. The apparatus is not a "stove," but resembles most the modern ovens used in connection with gasolene stoves; and the fuel used is common kerosene oil in an ordinary, or better still, a specially constructed kerosene lamp. Several articles of food may be cooked at the same time; bread may be baked; meat, vegetables, fish and poultry may be cooked; and all done so appetizingly and in such splendid condition for digestion as to leave nothing to be desired.

The leading principle in the cooker is the slow cooking, and the preservation of all the juices and flavor of the food—especially meats.

The stove (oven) may be set upon a table in the "sitting-room" with the breakfast for the following day in it. This may consist of quite a variety of food, and in quantity sufficient for from twelve to twenty persons. The husband may read his paper by the light of the lamp that does the cooking, and the wife may sew and rock the baby, and at bedtime the flame may be turned down partly; and in the morning there is no fire to make; no preparation for breakfast; nothing but to "set the table" and lift the victuals—and such victuals! the most capricious stomach couldn't resist the impulse to eat them. Full directions accompany the stove, relative to the time required for different articles of food to be properly cooked. Nothing can be burned—nothing can be made crisp; everything is tender and juicy and nutritious.

This is no fancy sketch; no extravagant claim made by the inventor for advertising purposes, because we have seen the cooking done and eaten the food cooked by it, and know whereof we speak.

We have not attempted a detailed description of it—we have simply suggested its form and appearance. It must be supplemented by some other means of heating for other purposes. It will not boil a teakettle. It will not do for a laundry stove; nor for making soap. It will not do to warm by. It is essentially a *cooker*—emphatically *the* cooker!

BOSTON, Oct. 21, 1891.

MR. EDWARD ATKINSON.

Dear Sir:—I have heard of your oven for several years, and have heard nothing but good, so I expected to find a good cooker, but I had no idea how much labor is saved. I believe the fatigue to be not more than a quarter what it is in using a range or stove, and feel that I have nothing to do, though I do all the cooking for a family of four.

After a few trials, one is able to put the food into the oven and leave until cooked without moving it, and there is no looking at the fire to see if is too hot or going out, and until one has tried one of your ovens, they don't realize how much time was wasted in watching food to keep it from burning.

I advise every family to add an Aladdin Oven to whatever cooking apparatus they now use, and feel assured they will use the other less and less, and in time, in many instances, do away with the stove or range. I believe that in an appartment, when hot water is provided, there is no occasion for any other than the oven, as irons can be heated on the bottom of it, as near as possible over the lamp; three irons can be kept hot at once.

I can go to church on Sunday, and be at home at 12.20, and have a dinner of roast beef, two or more vegetables and a pudding ready in two hours, and when the whole dinner is cleared away, I am not overheated or in the least tired by it. I believe the oven appeals to every intelligent housekeeper, whether they have servants or not. I also believe that if a servant was once in a house where one was used, she would never sit in an overheated kitchen where there was a range.

I have always used the best cuts of meat, and still do so, therefore the cost in food is not changed that I know of with us; but the cost of labor and fuel is enough to satisfy me, and the flavor of everything is improved. I use no other cooker excepting a small oil stove, and I could do without that very well. I make tea and coffee in the oven. I should add that I am in an apartment where hot water is provided.

I would not be without one of your ovens on any account, and wonder how I ever stood the heat of a range.

Yours truly,

Hotel Westland, Westland Avenue. J. P. MORSE.

Mrs. Edward Atkinson sends us her verdict in the following terms:

EDWARD ATKINSON.

You ask for my opinion of the Aladdin Oven. I think too much cannot be said in its praise. It has been an *immense* comfort and convenience ever since its introduction into our kitchen more than three years ago, and it has lightened the burden of housekeeping to an almost incredible extent. It really *is* as if a magician waved his wand, when a Thanksgiving dinner for twenty-four persons can be cooked with but a feather's weight of the care and trouble necessary in the use of the range or cooking stove.

Yet the idea of *quickness* associated with the feats of a magician must be discarded in the use of the Aladdin Oven. For although frying may be imitated in this oven with comparative speed, *slowness* in cooking is the grand desideratum— and also the great merit of the oven.

MARY C. ATKINSON.

Brookline, Jan. 13, 1891.

Finally, the question may be asked, What do the cooks say about this apparatus? Those who are intelligent and not opinionated are glad to adopt this oven for many reasons: it saves a large part of the work and renders it wholly unnecessary to stand over a hot stove while attending to the preparation of food. The better way will be for the stenographer to whom this statement is being dictated to get from the cook, who for over two years has prepared a mid-day meal for the employees, numbering over twenty, in an office in Boston, at an average cost of not exceeding twenty cents a day for a substantial and excellent meal, a statement of her experience in her own words without any change or variation by the writer. I gave her a few verbal instructions only. Her statement is as follows:

"When I first undertook to cook in this way I knew nothing about the method, and had never even used a kerosene stove of any kind whatever. But I thought if any one could succeed, I could, so I was not afraid to try. I have liked the ovens from the beginning, finding my work easier after the first preparation, nearly one-half the time of the work being saved by not having to watch and tend it; and I also find the saving of heat and the necessary work connected with a common stove is very great. I have never had the slightest failure or waste in the cooking of the food from the time I first undertook it. I can do everything except fry, and I can imitate a fry or a broil very nicely. With several ovens, an outside supply of hot water, and a three-holed iron table for the making of sauces, gravies, tea and coffee, etc., with one assistant I could do the cooking for one hundred people. I use one of the smaller ovens at my home, where I do the cooking for my own family of three, at night after I return home from my day's work."

It should be observed that the room in which this woman does her work is on the middle floor of a large office building. It was intended to be let for an office; it has no chimney-flue in it. It is ventilated by the outer windows and by a transom window over the door into the hallway. The adjoining room is used as a dining-room. The landlord permitted the use of this office for a kitchen on condition that the work should be stopped at a day's notice if there were any serious complaint. There has been none of any moment, although in the first experiments the fact could be detected that cooking was going on somewhere in the building, by a little odor. For the last year or more there has been no suspicion attached to the room; and a new tenant who occupied the room next to the dining-room, next but one to this kitchen, did not find out for three weeks that there was any cooking going on near him.

JOLIET, ILL., April 17, 1891.

Dear Sir :—For several days past I have had part of the sponge baked in the Aladdin Oven and part in the stove, with the following results in every instance: When baked in the stove the bread raises well, is very white, close texture, but has not much taste. When baked in the Aladdin Oven the bread is more yellow, somewhat coarser, raises well, and is very sweet and palatable, and keeps moist longer—in short, is preferable. Do you have the same experience?

Yours truly,

L. H. HYDE.

JOLIET, ILL., March 4, 1891.

Dear Sir :—I will ship the barrel of flour next week. With your permission I will not ship a full barrel of flour, but will put a separate package in the top of the barrel of what I call "Germ Graham," believing it will please you.

In reducing wheat to flour, the grain, after being cleaned, passes through a pair of rolls that reduce it to large pieces of bran with flour clinging to them, middlings of various sizes and some flour; it then passes to a sieve scalper that separates the coarser pieces from the middlings and flour. This coarse product then goes to another pair of rolls and through the same process again until five pair of rolls, or five reductions in all have been employed. The middlings thus made are separated from the flour, and then graded according to size. The coarsest middlings are called the germ middlings and contain the germ particles of bran, especially the inner or nitrogenous coat clinging to the middlings and the large glutinous middlings, only no starch, as the rough treatment they have so far been subjected to has pulverized all the starch (which is brittle) to flour; the gluten, being more tenacious, has retained the shape of irregular chunks. These germ middlings are run through a pair of smooth rolls, and reduced in size and partially flattened. The particles of germ being oily are reduced to little flat cakes. This product contains, in a condensed form, the oily germ (the sweet part of the wheat), the best of the gluten, the nitrogenous part of the bran and is free from the starch and coarse bran. It is condensed because it would take a number of barrels of ordinary flour to contain an equal amount of germ, gluten and nitrogen as is contained in one barrel of this. It is better than ordinary graham flour, because it does not contain the starch and woody fibre of the bran, and is made out of Minnesota Fyfee wheat. It makes a very delicious brown bread that will keep moist and sweet.

I will also send you a small sample of the germ middlings before they pass to the roll ; and will answer your letter more fully then.

Yours truly,

Louis H. Hyde.

AFFAIRS OF THE HOUSEHOLD.

[Evening Post, New York, September 21, 1891.]

An intelligent woman who has given the Aladdin Oven a thorough trial, says of it : "I confess it is almost a disadvantage that those who know this oven by actual use are able to claim so much for it, for it sounds like magic, as the night cooking of soup, etc., and makes persons unacquainted with it incredulous of everything said concerning it, but I think I shall not claim too much if I give a list of the advantages which I have found in six months' use of it, beginning with the lesser benefits and going on to the greater:

"(1.) Economy of fuel, saving seventy-five per cent.

"(2.) Economy of cooking utensils, saving ninety-five per cent.

"(3.) Economy of dish and glass towels, as they cannot be scorched and burned up handling red-hot vessels.

"(4.) Economy in annoyance and mortification in finding the house full of odors, and subsequent waste of flavor in dishes on the table.

"(5.) Economy in labor, saving fifty per cent. to the cook.

"(6.) Economy in time, saving fifty per cent. to the cook. No attention required.

"(7.) Economy in force, saving exhaustion from overheating, as there is no radiation of heat as from a range.

" (8.) Economy of skill, since a poor cook is not obliged to use skill in regulating the heat.

" (9) Economy of food spoiled by cooking, saving 100 per cent., and dishes wholly unlike may be cooked at the same time, and no flavor is imparted from one to the other.

"(10.) Economy in marketing, as inferior pieces of meat are rendered palatable and attractive by this method.

"(11.) Economy in conserving the nutritive qualities of food material in the highest degree known."

In the face of such evidence as this, it is wonderful that any housewife should hesitate about adopting an invention more necessary to a well-conducted household than a clothes-wringer, a carpet-sweeper, or a sewing machine.

In the New England Diet Kitchen of Boston, before referred to in this column, the Aladdin Ovens are also most successfully used in the valuable experiments in in scientific cookery.

In the first report published of this work, which is under the direction of the Massachusetts Institute of Technology, the American cook stove is condemned as "a crude and unreliable means" of preparing food. The report says: "As we have demonstrated that for the most substances a long and moderate heat is required for the best results, we must find an apparatus that secures this.

"No hotel or restaurant have we found in which the apparatus is made with regard to definite scientific principles.

"For ordinary cooking apparatus, these are some of the essential points:

"The degree of heat should be under perfect control, increased, diminished, or withdrawn at will, and without loss of time. This can only be perfectly attained with liquid or gaseous fuel.

"The use of a close oven, with a non-evaporative atmosphere, seems to be the secret of the retention of the delicate and volatile flavors which usually pervade the house and street, and not the food when it is brought to the table."

GENERAL INDEX.

	PAGE
Abel, Mrs. Mary Hinman, Report of,	136
Advertisement, How and How not,	10
Amusement in the Food Fad,	110
Art of Cooking, Pop. Science Monthly,	3
Author, Grotesque aspect of,	109
Baking, Aladdin Oven,	72
Beans, Aladdin Oven,	73
Beans and Peas, India Varieties,	9
Bean, Soja or Soy,	8
Bills of Fare, Mrs. E. H. Richards,	105
Bills of Fare, Specific,	59
Body, Uses of Food,	45
Box Cooking,	135
Boy, Office, Cooking in one lesson,	28
Braising, Aladdin Oven,	72
Bread, Aladdin Oven, Directions for,	71
Bread, How to make,	27
Bread, Prices of,	17
Bread-Kneader, Described,	27
Bread-Raiser, Description of,	27
Bread-Raiser, Where to get one,	71
Breakfast for perverse flies,	107
Broiling, Aladdin Oven,	72
Broth, Beef, N. E. Kitchen,	76
Calory, Scientific measure,	34
Calory, Unit of capacity,	16
Calories, Definition of,	15
Calories, Variable cost of,	56
Cake, Aladdin Oven,	72
Carnegie, Andrew, Contribution of,	26
Cents, Twenty-five, Nutrient purchase,	58
Children, How to attract,	1
Church, Prof., London Dietary,	125
Church, Prof., on Rice,	8

	PAGE
Claims of Writer,	31
Coal, Waste of,	13
Cod Liver Oil, Substitute,	101
Coffee, Southern Method,	78
Columbia College Lecture,	11
Comparative Nutrition,	147
Constants, Definition of,	112
Cook, Plain, How to Make One,	6
Cookery Books, Deficiencies,	28
Cookery Books, Merits and Defects,	28
Cooking, Art of, Definition,	1
Cooking, Box,	135
Cooking, Good, What is it?	30
Cooking, Slow, Merits of,	129
Cooking, Slow, Mrs. Abel,	136
Cooking, True Object of,	28
Cooking, Two Simple Rules for,	5
Cooking, Very Slow,	76
Corpulence, Dietaries for, Atwater,	36
Cost per Week, Dietaries,	125
Daniell Miss Maria, Report,	127
Dietaries, By Classes, Graphical, Atwater,	38
Dietaries, By Classes, Atwater,	57
Dietaries, Definition of,	18
Dietaries, For Corpulence, Atwater,	36
Dietaries, For Thirty Days,	110
Dietaries, Specific by Days,	59
Dietaries, Thirty Days, Specific,	113-124
Digestion, Nutrients, Proportion of,	45
Digestibility, Food Varieties of,	44
Dinner, An Aladdin Oven,	73
Dinner, For Twelve at Sixty-one Cents,	22
Dinner, Seven Course, Thirteen cents per Head,	22

GENERAL INDEX.

Dwelling, How to Enlarge,	26	Gas, How to Use with Oven,	70
Eggs, Consumption of,	23	Gumption, Proportion in each dish,	27
Egleston, Prof., Authority of,	11	Havemeyer, T. A., Contribution of,	26
Emmet, Dr. T. A., Pork for Cod Liver Oil,	101	Health, Good Evidences of,	61
		Heat, Function of,	30
Energy, The Starting-point,	61	Income, Per cent. for Food,	47
Expectations, Not Warranted,	30	Inertia of Woman, To be Overcome,	11
Extravagance In Food,	46	Kitchen, Aladdin,	106
Feeding, Bad, What it Costs,	25	Kitchen, New England,	76
Five Hour Cooking,	127	Kitchen, New England, Story of,	65
Fleishmann's Yeast,	85	Kitchens, People's,	63
Flour, Description of,	79	Laboratory Cooking,	106
Fly Escape, Described,	107	Laboratory Fittings,	107
Food and Wages, Relation of,	33	Lamp, Care of, Mrs. Abel,	137
Food Bill, National,	.	Lamps, Care of,	104
Food Bill, National,	25	Lamps, Description of,	68
Food, Comparative Expense of, Graphical,	51-52	Liquors, Cost of,	23
		Mealers, Lowell Price of Meals,	23
Foods, Composition of, Graphical, Mrs. E. H. Richards,	64-65	Meals, Fast Day, Cheap,	20-21
		Mission of the Author,	11
Food Concentration, Sir Henry Thompson, M. D.,	22	Mistress Roasted, No Longer,	102
		Nitrogen, Pulse as Source,	9
Food, Cost of to Multitude,	5	Nutrition, American Standard,	18
Food, Daily need of, Atwater,	35	Nutrition, Beasts Known, Men Almost Unknown,	14
Foods, Digestibility of,	44		
Foods, Economical,	50	Nutrition, Cost of,	19
Foods, Expensive,	50	Nutrition, Science of,	109
Food, Extravagance in,	46	Nutrition, Standard of,	15
Food, Half the Cost of Living,	12	Nutrition, Standards Exact,	16
Food Material, Composition of, Atwater,	39	Nutrition, Standard of, Compared,	19
Food Materials, Per cent. of Nutrients,	42-43	Nutrients, By Classes,	40
Food of Nations, Compared,	33	Nutrients, Digested and Not Digested,	45
Food, Potential Energy,	41	Nutrients, Function of,	14-15
Foods, Potential Energy, Graphical	55	Nutrients, Per Centage in Foods,	42-43
Food Power, Described,	33	Oat Meal, Merits of,	54
Food Preparations, Wholesale,	63	Oven, Aladdin, What it is,	66
Food, Preparation of,	27	Oven, Aladdin, What it does,	67
Food, Standard by Classes, Atwater,	36	Oven, Aladdin, How it does it,	78
Food, Uses in Body,	45	Oven, Aladdin, Mrs. Abel's Judgment,	137
Food, Wholesale Preparation in Boston,	64	Oven, Tests of Miss Parloa,	120
Fruit, Preserving without Sugar,	99	Oven, Aladdin, Merits of,	103
Frying Pan, American D—d,	30	Oven, Aladdin, Time Table,	130
Frying, Sir Henry Thompson, M. D.,	91	Oven, Hot Water Kind,	29
Fuel, Gaseous and Potential of,	13	Oven, Temperature,	137
Game, Aladdin Oven,	73	Pails, Cooking, Capacity of,	12

GENERAL INDEX. iii

	PAGE
Papin, Digester, What He did,	6
Pastry, Aladdin Oven,	72
Pea Cow, Renovator of Soil,	8
do. For Ensilage with Corn,	
Phipps, Mr., Contribution of,	26
Pierce, Henry L., Contribution of,	65
Pork Fat, Substitute for Cod Liver Oil,	101
Potato Gospel, Definition of,	14
Pulse, Necessary with Rice,	8
Pump, California, Inventors Compared,	7
Recipes, Aladdin Oven, General,	79
Rice, Varieties and Importance,	8
Roasting, Aladdin Oven,	72
Rumford, Count, What He did,	6
Rumford, Count, What He didn't. Why?	6
Ration, Life,	101
Ration, Work,	114
Sauces, How to Make,	75
Sauces, Importance of,	29
Sautéing, Aladdin Oven,	79
Science of Cooking Defined,	4

	PAGE
Scolds, Swear Words,	66
Simmering, Aladdin Oven,	72
Soup Stock, Aladdin Oven,	72
Statistics, Distrust of,	12
Sterling, Mrs., Time Tables,	168
Stove, Iron, Infernal Machine,	7
Summer Cooking, No Coal Needed,	13
Thermometers, Cooking and Substitute for.	2
Time Table, Aladdin Oven,	130
Variables, Definition of,	112
Vegetable Cooking of Aladdin Oven,	72
Vegetarian, Dietary,	120
Waste, How to Measure,	24
Water Heating,	103
Water Warming,	71
Watts, Dr. George, Food Products of India,	8
Williams, Dr. M., Cheese Cooking,	83
Women in Shops, How They Starve,	29
Workman's Cooking Pail,	103

INDEX TO RECIPES.

Those marked "E. A." have been developed by sundry persons in practice under the direction of the Author; those marked "M. D." are furnished by Miss Maria Daniell; those marked "M. H. A." are furnished by Mrs. Mary H. Abel.

	PAGE		PAGE
All Sort of Things, Edward Atkinson,	91	Biscuits, E. A.,	80
Asparagus, E. A.,	95	Biscuits, Rye, E. A.,	97
Bacon, E. A.,	80	Cake, Spice, E. A.,	97
Baked Beans, E. A.,	82	Cake, Sponge, E. A.,	97
Beans Stewed, Maria Daniell,	132	Cake, Thin Indian, E. A.,	98
Beef a la Mode, M. D.,	135	Calf's Heart, M. D.,	132
Beef a la Mode, M. D.,	131	Celery, Stewed, E. A.,	95
Beef Fricassee, M. D.,	133	Cheese, E. A.,	83
Beef Gravy, E. A.,	94	Cheese Dishes, M. H. A.,	142
Beef and Oatmeal, M. D.,	130	Chicken, Baked, E. A.,	94
Beef Roll, M. D.,	131	Chicken, Broiled, E. A.,	79
Beef Skirt, M. D.,	131	Chicken, Broiled, E. A.,	94
Beef Spiced, E. A.,	95	Chicken, Potted, E. A.,	98
Beef Steak, E. A.,	79	Chicken, With Sauces, E. A.,	86
Beef Steak, E. A.,	94	Chowder, E. A.,	82
Beef Stew, M. D.,	131	Clams, Seaside Fashion,	82
Birds, Aladdin, E. A.,	98	Coffee, E. A.,	97
Boiled Dish, E. A.,	99	Consommé, M. D.,	135
Box Cooking, E. A.,	135	Corned Beef, E. A.,	94
Bread, E. A.,	84	Cream Sauce, E. A.,	94
Bread, M. D.,	134	Crispy Cake, E. A.,	88
Bread Baking, Mary H. Abel,	138	Dinners, Specific, M. H. A.,	143, 144
Bread, Brown, E. A.,	97	Duck, E. A.,	87
Bread, Case's Health,	84	Eggs, M. H. A.,	141
Bread, Corn, E. A.,	97	Eggs, E. A.,	80
Bread, Graham, E. A.,	97	Fads, Social, Described,	31
Bread and Raisin Cake, E. A.,	90	Fats, Noxious Effect of High-heat,	29
Bread, Sir Henry Thompson, M. D.,	85	Fish, M. H. A.,	140
Bread, White, E. A.,	97	Fish, E. A.,	80
Brown Bread, E. A.,	82	Fish, Seaside Fashion, E. A.,	82
Browning Fish,	81	Flies, Breakfast, E. A.,	107
Browning Meat, E. A.,	81	Fowls, E. A.,	86

INDEX TO RECIPES.

	PAGE		PAGE
Fruit, Drying, M. H. A.,	146	Pork Fat, E. A.,	101
Fruit, M. H. A.,	143	Porridge, E. A.,	80
Fruit, Preserving, E. A.,	90	Poultry, Aladdin, E. A.,	95
Gander, Minced, E. A.,	86	Puddings, M. H. A.,	143
Game, Potted, E. A.,	98	Pudding, Apple,	95
Gingerbread, E. A.,	97	Pudding, Batter, E. A.,	96
Gingerbread, E. A.,	88	Pudding, Birds' Nest, E. A.,	96
Glaze, M. D.,	134	Puddings, Box, E. A.,	80
Grains, M. H. A.,	139	Pudding, Brown, E. A.,	96
Grains, M. D.,	133	Pudding, Brown, E. A.,	88
Griddle Cakes, E. A.,	80	Pudding, Cottage,	96
Grouse, E. A.,	87	Pudding, Cracker, E. A.,	96
Grouse, E. A.,	94	Pudding, Ground Rice, E. A.,	96
Haddock and Tomato, M. D.,	138	Pudding, Hasty, E. A.,	96
Haddock and Tomato, M. D.,	132	Pudding, Indian, M. D.,	133
Halibut, a la crême, E. A.,	95	Pudding, Indian, E. A.,	96
Ham, E. A.,	80	Pudding, Oatmeal, M. D.,	133
Hams, Whole, E. A.,	83	Pudding, Pease, M. D.,	132
Hash, Sam Weller, E. A.,	82	Pudding, Plum, M. D.,	133
"Hog and Hominy," E. A.,	83	Pudding, Poor Man's, E. A.,	96
Hominy, E. A.,	98	Puddings, Thanksgiving, M. D.,	133
Indian Pudding, E. A.,	82	Quail, E. A.,	95
Macaroni, E. A.,	87	Quail, with Sauces, E. A.,	86
Meats, M. H. A.,	141	Roast Beef, E. A.,	94
Mushes, E. A.,	80	Salt Cod, M. D.,	132
Mutton Chop, E. A.,	79	Salt Fish, E. A.,	83
Mutton Chop, E. A.,	94	Sausages, E. A.,	80
Mutton Stew, M. D.,	132	Sausage Stewed, M. D.,	131
Mutton and Tomato, M. D.,	132	Sheep's Heart, M. D.,	132
Omelet, E. A.,	81	Shrewsbury Cake, E. A.,	89
Omelet, E. A.,	80	Soda Biscuits, M. H. A.,	139
Oyster Plant, E. A.,	95	Soups, E. A.,	93
Pan Cakes, E. A.,	81	Soup Stock, M. H. A.,	140
Parker House Rolls, E. A.,	88	Soup, Shin of Beef, E. A.,	93
Partridge, with Sauces, E. A.,	86	Steak Pudding, M. D.,	132
Pastry, E. A.,	84	Stock, M. D.,	134
Pease Pudding, M. D.,	132	Toasting Bread, E. A.,	81
Pie Crust, E. A.,	99	Tripe, M. D.,	130
Pies, M. H. A.,	143	Veal, E. A.,	94
Pigs' Feet, M. D.,	130	Veal Cutlets, E. A.,	80
Pop Overs, E. A.,	96	Vegetables, M. H. A.,	142
Pork and Apples, E. A.,	94	Vegetable Soup, M. H. A.,	140
Pork Chops, E. A.,	80	Water Heater, M. H. A.,	143
Pork, Corned, M. D.,	132	Wheat Cakes, Roasted, E. A.,	81

www.ingramcontent.com/pod-product-compliance
Lightning Source LLC
Chambersburg PA
CBHW020243170426
43202CB00008B/201